Walking Machines: An Introduction to Legged Robots

WALKING MACHINES
AN INTRODUCTION TO LEGGED ROBOTS

D J Todd

Kogan Page

First published in 1985 by Kogan Page Ltd
120 Pentonville Road, London N1 9JN

Acknowledgements
I should like to thank all those who responded to my requests for illustrations, and in particular
Professors Raibert and Hirose for their expressions of interest. I am indebted to Mr D C Witt
for his demonstration of walking aids and information on early research on locomotion. I
should also like to thank Kogan Page Ltd, and in particular Kevin White, for their diligent
efforts in the editing and production of this book.

British Library Cataloguing in Publication Data
Todd, D. J.
    Walking machines: an introduction to legged robots.
    1. Robots
    I. Title
    629.8'92   TJ211

    ISBN 0-85038-932-1

Printed in Great Britain by Anchor Press Ltd
and bound by William Brendon & Son Ltd
both of Tiptree, Essex

Contents

Preface

The first chapter of this book traces the history of the development of walking machines from the original ideas of man-amplifiers and military rough-ground transport to today's diverse academic and industrial research and development projects. It concludes with a brief account of research on other unusual methods of locomotion.

The heart of the book is the next three chapters on the theory and engineering of legged robots. Chapter 2 presents the basics of land locomotion, going on to consider the energetics of legged movement and the description and classification of gaits. Chapter 3, dealing with the mechanics of legged vehicles, goes into leg number and arrangement, and discusses mechanical design and actuation methods. Chapter 4 deals with analysis and control, describing the aims of control theory and the methods of modelling and control which have been used for both highly dynamic robots and multi-legged machines.

Having dealt with the theory of control it is necessary to discuss the computing system on which control is to be implemented. This is done in Chapter 5, which covers architectures, sensing, algorithms and programming languages. Chapter 6 brings together the threads of the theory and engineering discussed in earlier chapters and summarizes the current walking machine research projects. Finally, the applications, both actual and potential, of legged locomotion are described.

Introduction

Research into legged machines is expanding rapidly. There are several reasons why this is happening at this particular time. The main one is that it has recently become practicable to build on-board computers into small vehicles. Without computer control it is almost impossible to coordinate the movement of many joints as is needed for smooth motion on a rough and obstacle-littered surface. In the last few years the development of computer-controlled machines, especially industrial robots, has resulted in techniques which taken together provide most of the technical base needed to make walking machines possible.

Also, the increasing familiarity of industrial robots and automatically guided vehicles has rendered the idea of new kinds of robot more respectable: anyone who has seen a film of a spot-welding robot in action can see that there should be no insuperable obstacles to making a robot with legs instead of arms.

The final reason why legged robot research is taking off is that new applications are being found. For many years walking machines have been thought of as rough-country vehicles, but recently a new impetus has come from the expansion of the need to handle radioactive materials in awkward environments and, in particular, the repair and decommissioning of nuclear power stations. Other potential applications are also starting to appear, as described in Chapter 7.

Advantages of Legs

What is the case for building legged vehicles rather than wheeled or tracked ones? One reason is an interest in legged locomotion in itself, but where the purpose is transport, legs must be shown to be superior to wheels or tracks. The advantages of legs can be summarized as follows:

1) Legs can step over obstacles and up and down stairs.
2) Legged locomotion can, in principle, even carry a vehicle over wide chasms or extremely broken ground (consider kangaroos and mountain goats). Such performance is not expected in the near future.
3) A legged vehicle can achieve a smooth ride on rough ground by

varying the effective length of its legs to match the undulations of the ground.

4) On soft ground a wheel is always climbing out of a rut of its own making; this wastes power. In an extreme case the wheel may just dig itself deeper until the vehicle stops.

5) Legs do less damage to the ground than tracks and many wheels.

The comparative merits of walking and other forms of location are discussed in Chapter 2.

Terminology

Throughout this book the terms 'robot' and 'vehicle' are used somewhat interchangeably (as are 'legged' and 'walking', although a legged entity may use modes of locomotion other than walking). The term 'mobile' will also be used as a noun, to denote a self-propelled entity without distinguishing between animals and machines.

Although it is not strictly adhered to in this book, it is necessary perhaps to give a definition of a robot, as it is not a word which can be used in the knowledge that there is general agreement on its meaning.

1) A robot must be produced by manufacture rather than biology (this does not rule out the eventual use of artificial biochemically produced structures such as muscles).

2) It must be able to move physical objects or be mobile itself (this excludes simulations and control systems for static plant).

3) It must be a power or force source or amplifier (this excludes those teleoperator arms which merely replicate an operator's hand movements).

4) It must be capable of some sustained action without intervention by an external agent (this rules out ordinary road vehicles).

5) It must be able to modify its behaviour in response to sensed properties of its environment, and therefore must be equipped with sensors of some kind.

A more informal view of a robot, which encompasses most forms met so far, is a machine which possesses functional arms or legs, or else is a driverless vehicle. This allows the inclusion of the simpler types of industrial robot which would be excluded by the narrower definition.

There is scope for argument in certain cases. For example, is a legged vehicle with a human driver a robot or not? It would seem that General Electric's Walking Truck, in which the driver had to directly command every joint movement, was not, but that a machine which can walk automatically but whose speed and direction are commanded by its driver does fall within the definition (just). The same mechanism with an automatic navigator instead of a driver would definitely be a robot.

Chapter 1

A brief history of walking machines

The earliest walking machines were mechanical toys. (The history of automata and mechanical toys is outside the scope of this book.) Their legs were driven by cranks or cams from a source of rotary power, usually clockwork, and executed a fixed cycle. An interesting variant is the walking toy which is not internally powered but can walk down a slight slope under gravity power, the legs being independently pivoted and not driven by cranks. This principle has been adopted in some prosthetic walking aids. It may have some lessons for the design of power-economical robots.

In passing, the steam elephant in one of Jules Verne's novels warrants attention, as this seems to anticipate one of the most successful methods in robotics, namely the use of pneumatic or hydraulic actuation of individual joints.

The main event which led to the consideration of practical walking machines was the invention of the internal combustion engine, which made possible all sorts of new vehicles. One line of development consisted of extending wheels and caterpillar tracks by attaching feet, spikes or posts to the rim or to sections of the track. Several 'legged' machines of this kind are described by Thring (1983). However, these might not be considered true walking machines as they do not have legs separately attached to the body.

Another line of development has been the production of giant excavators such as walking draglines. An example is the Ransomes & Rapier walking dragline which, in 1962, held the world weight record at 1800 tons. While excavating it sits on a large flat circular base. To move (its speed is $1/8$ miles/hour), it drives its two flat feet, one on each side and each the size of a large bulldozer, until they take enough weight to drag the machine backwards. However, such excavators provide few lessons for the design of more agile vehicles.

A third branch of mobile robot development has been the invention of machines to move in specific environments, such as within pipes, up trees or on orbiting structures. Some of these are mentioned in Chapter 7.

It is possible to identify two principal themes of legged robot research and some subsidiary ones. The two main themes are rough-terrain transport, and prosthetic or orthotic bipeds. There is not an absolute distinction between these. The bipedal exoskeleton worn by a person has been seen not just as an aid for the disabled but as a strength amplifier, and one application of this is in transport. The General Electric Walking Truck (described later) is a hybrid, in that it walked as a quadruped but was controlled as a sort of exoskeleton attached to its driver. Of the subsidiary themes perhaps the most important has been the study of animal locomotion. In the beginning most research was aimed at military transport and prostheses; however, there have been many developments — as described in Chapter 7.

Rough-Terrain Transport

The earliest serious attempt to build a walking machine with independently controlled legs was made in Britain in 1940 by A. C. Hutchinson and F. S. Smith (Hutchinson 1967). Hutchinson, who worked for W. H. Allen & Company Ltd, suggested that for a very large armoured vehicle, in the 1000 ton class, legs would be better than tracks. Hutchinson and Smith decided on four legs and the quadruped crawl gait. They studied a variety of leg mechanisms, mainly intended to allow a pair of hydraulic actuators to produce easily separable vertical and horizontal movement. They chose a design with a rolling thigh joint, which acted as a kind of inverted wheel, and a telescoping leg.

The proposed control mechanism consisted of a feedback loop per leg, with the four legs each being controlled by the hand or foot of the driver: essentially the solution to be adopted by General Electric for the Walking Truck in the 1960s.

Hutchinson and Smith built a four legged model, about 60cm high, whose eight joints were controlled by flexible cables which led to a console at which the operator sat, with his feet on the two hind-leg pedals and his hands on the two fore-leg handles. The model was made to walk and to climb over a pile of books. (Not surprisingly, in 1940 the UK War Department was not prepared to abandon its commitment to the well established tracked tank, and the project was halted.)

The main stream of modern legged vehicle research begins with a group associated with the University of Michigan and the US Army Tank-Automotive Center (ATAC) at Warren, Michigan (sometimes referred to as the Ordnance Tank-Automotive Center or OTAC). The US Army has maintained an interest in machines for rough-ground transport since the Second World War. The Army (or sometimes the Defence Advanced Research Projects Agency, DARPA) has sponsored much of the American research on legged vehicles. The group started with M. G. Bekker, who founded ATAC's Land Locomotion Laboratory in the early 1950s. Bekker has been studying land locomotion for many years and has written several books on the subject. The work of the laboratory was continued by R. A. Liston, who wrote a much referred-to paper on walking machines in *Terramechanics Journal* (Liston 1964). Another associate was R. K. Bernhard of Rutgers University, who in 1954 began an investigation of gaits for legged vehicles. In a 1957 study he examined several types of levered vehicle, including leaping, galloping, bouncing and running machines. His prime criterion was vehicle speed, but most high speed schemes presented difficult control and vibration problems, and the most tractable seemed to be running. This was followed up by a further study, with J. E. Shigley of the University of Michigan, who had worked on designs for legged tanks (Shigley 1961; Hutchinson 1967). This study established a set of criteria for an ideal walking machine. These criteria covered the relative dimensions of the machine, the quantities to be controlled and ways of achieving smooth and efficient walking.

Shigley designed a walking machine with all-mechanical linkages which was intended to fulfil most of these requirements, and which was subsequently built. The main problem was its use of non-circular gears which reduced the accelerations of the legs but were difficult to make. The performance of the vehicle was limited and it seemed that a hydraulic system offered more potential, so it was abandoned. The hydraulic design used a pantograph linkage to transfer the motion of a horizontal and a vertical actuator, both body mounted, to the foot. The proposed design had 16 legs, a 'gang' of four at each corner, and satisfied Shigley's criteria. Unfortunately, no adequate way of controlling such a machine on rough ground and among obstacles was found and it too was abandoned.

The Land Locomotion Laboratory evaluated various designs of other inventors. An interesting example is the machine proposed by Scruggs (see Figure 1.1). This had eight legs which could slide vertically and longitudinally. The horizontal slide of the four end legs can be rotated in the horizontal plane for steering. This design was rejected by ATAC, but is of interest because it anticipates the design of the Komatsu underwater robot described in Chapter 7.

In 1962 the laboratory began working with General Electric on what was to be an influential project. By this time Liston and Shigley had more or less come to a dead end because of the control problem (Liston 1966), when the laboratory was approached by H. Aurand of General Electric. General Electric had already built remote (telechiric) manipulators using force feedback to the operator's arms. The most famous of these 'cybernetic anthropomorphous machines' or CAMs as General Electric called them, was the Handiman (of which there is a well-known photograph of it being used to whirl a hula-hoop).

Aurand proposed using the same force feedback servo principle to make a bipedal walking machine. It was intended to be much larger than its human driver, so it was not an exoskeleton (see Figure 1.2).

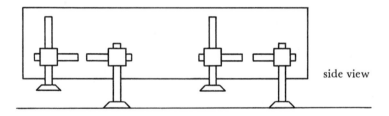

side view

Figure 1.1 *The Scruggs Walking Machine. In this side view the machine, which is walking to the left, stands on the second and fourth leg pairs and is about to lower the first and third pairs.*

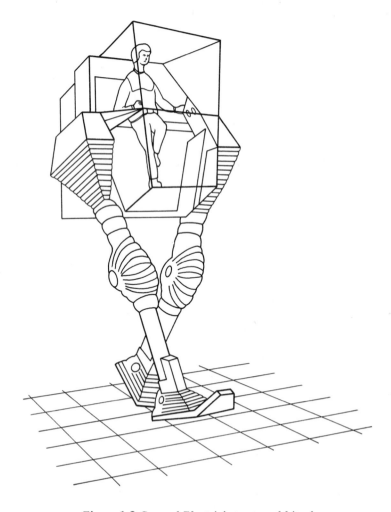

Figure 1.2 *General Electric's proposed biped.*

ATAC undertook a contract from General Electric to study the design of such machines; a balancing test rig was built and special General Electric servo valves incorporated. The study was judged successful and it was decided that a full-scale walking machine could be built. However, the geometry chosen as having more 'customer appeal' was a quadruped. In retrospect, perhaps this was a mistake, in that the operator might have found controlling the biped more tractable because it was more natural. If such a machine had succeeded, the subsequent history of walking machines might have been very different.

The quadruped actually built was a 'walking truck' designed by R. S. Mosher of General Electric's General Engineering Laboratory (Mosher 1968). Each leg was servo-controlled by the human driver, being coupled by hydraulic servo loops to an arm or leg, such that the truck's front legs followed the movements of the driver's arms, while the back legs followed those of the driver's own legs. The vehicle was over 3m long and weighed 1400kg (see Figure 1.3). Hydraulic power was generated by pumps driven by a 90 horse power petrol engine. The hand controllers for the front legs took the form of pistol grips, to which force feedback was reflected by the joint servo loops. Each handle controlled knee flexure and two axes of hip rotation. The driver's leg movements at hip and knee were followed in a geometrically similar way by the machine's back legs.

The quadruped performed well, but proved extremely demanding on its driver, who could not manage to control it for more than a few minutes at a time. Its importance lay in showing that the mechanical engineering of legged machines was tractable; also, by being widely publicized it probably stimulated other research on robots.

A second series of walking machine developments on behalf of the US Army, and also of NASA which at this time was getting interested in the problems of mobility in space, on the Moon and other planets, was undertaken about this time by Aerojet General Corporation (Morrison 1968). At least three applications were envisaged: planetary exploration; a walking chair for the disabled; and military transport (Morrison 1968; Vukobratovic 1973). Variants of the basic vehicle had six and eight legs and executed an alternating tripod and an alternating tetrapod gait, respectively. In the military transport role, a train of these vehicles walking behind and steered by a soldier on foot was envisaged. The leading vehicle was powered and towed the rest of the train. In this form it was known as the Iron Mule Train (see Figure 1.4).

The basic design used a fixed leg cycle produced by a mechanical linkage of cams and levers. A typical unit was 2m long and weighed 150kg. It had a separate walking mechanism for the four legs on each side, so it could turn by driving the two sides differentially as does a tracked vehicle.

The Iron Mule Train was tested by driving one of its powered units

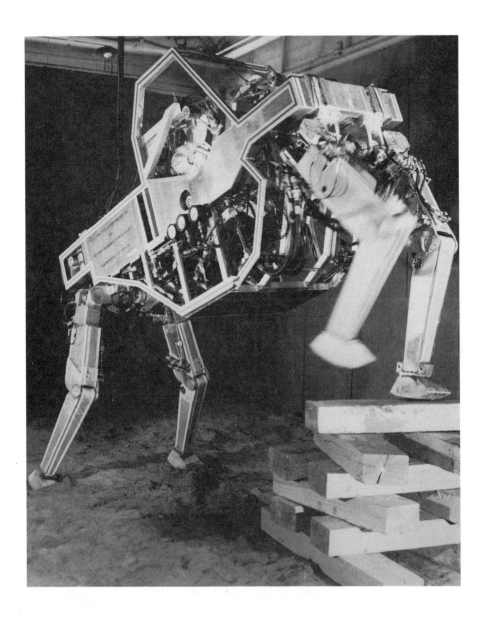

Figure 1.3 *The General Electric Walking Truck (courtesy US Army Tank Automotive Center).*

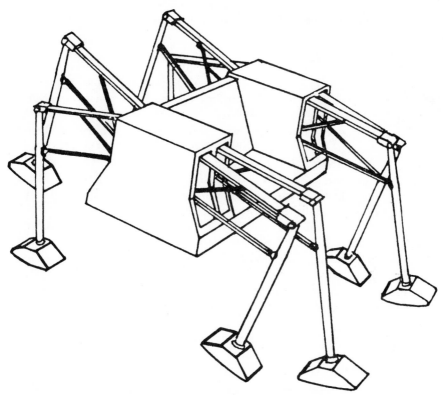

Figure 1.4 *The Iron Mule Train. The box at each side houses a set of cams, one cam for each knee drive and one for each hip. Each side has a single motor which drives all eight cams by a chain and sprockets.*

over a range of soils and measuring the conditions under which it failed. Unlike most subsequent walking machines it was tested in very rough conditions, for example dry soil with 0.3m deep potholes, or a 35° slope in loose soil. The conclusion was that it performed better than would have been expected of a comparable wheeled or tracked vehicle. However, this line of research was not pursued, presumably because the performance was still not good enough for the infantry supply application. This type of machine was also tested in its 'wheelchair' role. It could walk up and down stairs of only a limited range of profiles, and had difficulty in turning.

This was, perhaps, the last stand of the purely mechanically co-ordinated walking machine. Its lack of adaptability to ground conditions clearly outweighed the advantage of simplicity of design. The Iron Mule Train and the General Electric Walking Truck represented the last attempts to build walking machines without computers or electronic logic. The other main strand of early walking machine development will now be discussed, that is prosthetic legs and related research.

Bipeds: Medical Applications and Man-Amplifiers

The history of these machines emerges gradually from that of un-powered prosthetic and orthotic aids, which are extremely ancient in origin. It is also closely associated with the development of prosthetic arms. In addition to the medical application, exoskeletons were seen at one time as 'man-amplifiers', as discussed later. A third strand of this research has been the quest for more or less anthropomorphic bipedal robots. Although in principle this could be regarded as a separate subject, it has always been intimately linked with the prosthetic-orthotic application. This is at least partly because many potential users of bipedal aids, such as paraplegics, cannot exercise enough control over the mechanism for stability, so the control of walking has to be largely autonomous; in which case it is not a great step further to make the walking biped function without a human occupant at all.

It is uncertain when powered leg prostheses were first tried. In 1948 N. A. Bernstein in the USSR suggested a design for an above-knee prosthesis with powered knee joints, but this seems not to have had much immediate effect on research. One of the earliest practical efforts relating to bipeds was that of the Cornell Aeronautical Laboratory (CAL). In the mid-1950s Millikan and Eiken proposed the idea of a powered exoskeleton whose joints would be driven by servo mech-anisms duplicating the functions of the joints of the human body (Mizen 1968). It would be worn rather like a suit of armour. The man-amplifier, apart from its potential as an orthosis, was envisaged as a means of doing heavy work at remote sites not accessible to fork-lift trucks or cranes. It was supposed that although a person can perform only one or two simultaneous functions with a conventional machine, it would be possible by harnessing normal bodily movements to in effect carry out many controlled movements simultaneously. (This type of machine is of limited relevance to the problems of either prostheses or autonomous walking machines, but some aspects of design, such as those to do with the structure, power source and control of actuators, are common to all three areas.)

An unpowered exoskeleton was built as a design aid, but the CAL group then started to concentrate on arm exoskeletons and medical applications of these. General Electric built a high strength exoskeleton called Hardiman at the same period as the Walking Truck (see Figure 1.5). The actuation and servo control methods available seem not to have been adequate to make the exoskeleton a practical tool.

A research effort which more directly addressed the control issues of walking was that of D. C. Witt and associates at the University of Oxford between 1966 and 1978. The theme of this work was the development of practical aids for paraplegics and users with deformed or absent legs. The main approach eventually arrived at was to make use

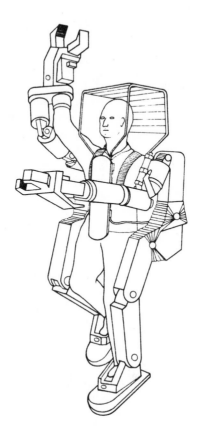

Figure 1.5 *The General Electric Hardiman (reproduced with permission from*
Robots and Telechirs: Manipulators with Memory; Remote Manipulators,
Machine Limbs for the Handicapped *by M. W. Thring published by*
Ellis Horwood Ltd, Chichester, England, 1983).

of the rider's ability to control and power the walker when possible.
Therefore, completely autonomous walking was not a primary goal.
However, in order to test various walking mechanisms, machines were
built which did walk autonomously.

The earlier designs were envisaged as being able to walk under their own
power from an on-board hydraulic supply. The control problem was decomposed into two, in the sagittal and frontal planes. A machine was
built, in 1968-71, to test the frontal plane control system. When supported only against falling backwards or forwards it could 'walk on the spot'
with satisfactory lateral stability and could adapt to a block placed under
one foot by shortening that leg to compensate (Hall and Witt 1971).

An interesting characteristic of the design was its intended use of only three actuators to achieve stable walking: one axial cylinder in each, telescopic, leg and one at the hip which would swing both legs in the sagittal plane only, through equal and opposite angles. In a way this anticipates the Carnegie-Mellon hopper's use of only three actuators to control stability in all planes. (However, control of the direction of walking seems not to have been considered in any detail.) This machine was successful as a first stage in achieving automatic locomotion, but Witt's primary aim was the development of practical aids, so it was abandoned in favour of passive walking aids. The basic design, shown in Figure 1.6, originated in observations of the movements by amelic (no legs) and phocomelic (short and deformed legs) children wearing swivel-walking or rocking aids. It used a four-bar, near parallelogram, linkage for each leg. The legs were coupled to each other and to the seat by levers and springs in such a way that rocking the body from side to side resulted in a forward (or if required backward) walking

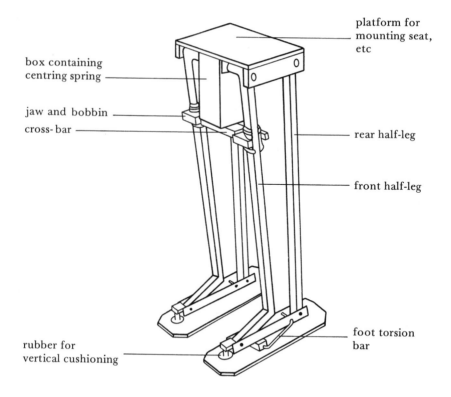

Figure 1.6 *A prototype of the Duckling walking aid (from Nichols and Witt 1971).*

motion (Nichols and Witt 1971). The walker worked quite well, but as is often the case with aids it was not used regularly because of factors unconnected with whether the aid meets its specification or not: in this case, its potential users thought it more fun to whizz about in an electric wheelchair than to shuffle slowly along on legs.

However, considered as a walking machine the device was of interest because when equipped with a powered 'dummy rider' for testing it became a self-contained walking machine of unusual design (see Figure 1.7). The dummy rider emulated the side-to-side rocking movement which a human rider would use to drive the walking mechanism. It did this by moving a mass (actually the hydraulic pump and reservoir), connected to the prosthesis by a linkage swinging in the frontal plane, using a hydraulic servo system. The machine walked satisfactorily on level ground, at a speed of 0.23m/s, compared with 0.33m/s for the human rider of a similar prosthesis. The power consumption (not taking electrical and hydraulic losses into account) was about 4 Watts.

A further aspect of this project was research on knee mechanisms which would allow a prosthetic walking machine to sit down, stand up and climb stairs. It was accepted that these activities would need intervention by the user by, for example, holding a handrail or removing locking pins, but it was required that the machine plus rider be statically stable throughout the standing or sitting operation (this was accomplished). The knee was hydraulically powered, and the movement of a roller over a cam ensured that the centre of gravity of the whole system moved in a near-vertical path over the feet.

An early interest in walking machines was also shown in Yugoslavia. In 1961 R. Tomovic of the University of Belgrade wrote a paper on finite-state control (Tomovic 1961) which influenced the development of walking machines in the US. This work was presumably known to Vukobratovic who subsequently did research on bipedal walking and prosthetics at the Mikhail Pupin Institute, Belgrade.

The Modern Era: Computer Control

From the 1960s onwards practically all legged machines, whether intended as autonomous robots, walking passenger vehicles or prosthetic aids, have used some degree of computer control. In a few cases this has been restricted to analog or hard-wired electronic logic circuits, but the desirability of flexible stored-program control has long been obvious, and the advent of, first, small minicomputers and then microprocessors has made it possible to build some sort of computing power into even the smallest of machines. In passing, it may be remarked that since most walking machines are ultimately controlled by a person it is necessary to decide how to apportion the control task between

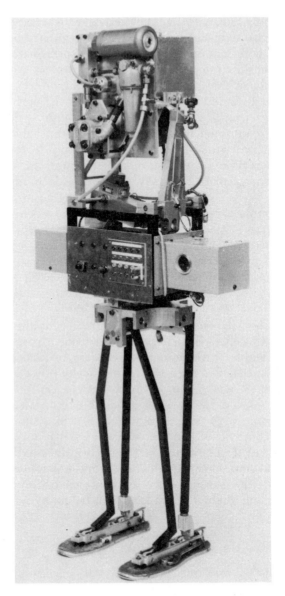

Figure 1.7 *The hydraulic Dummy Rider mounted on a walking aid (courtesy D. C. Witt).*

man and machine. In brief, it is common to structure the system into levels, with the lower levels, concerned with details such as servo control of joints, being handled by the computer(s), while the most strategic decisions are made by the person.

In the computer era the two main themes, those of many legged vehicles for rough-ground transport, and bipeds as prosthetic aids, are still discernible, although less distinct. Two main trends can be observed in the research of the 1970s and early 1980s. The first is research on improving the techniques for controlling machines with many legs so as to cope with more and more complex problems of terrain and obstacle adaptation. The second is the attempt to advance the concepts and techniques of dynamic control sufficiently to make it possible to build machines which can execute forms of locomotion requiring highly developed control of balance, such as running and jumping.

There has been a strong theoretical component in the research on legged locomotion and this work parallels and merges with research on natural locomotion in animals and people. (This book does not attempt to deal with natural locomotion except in so far as it is relevant to the design of machines.)

This section describes the first generation of computer-controlled walking robots. Some of these still exist but are now used as tools for developing terrain adaptability and other techniques, so a rather arbitrary line will have to be drawn between this chapter and the one on current research (see Chapter 6). It begins with what was probably the first legged machine whose walking cycle was generated by a digital electronic controller, the Phony Pony, developed at the University of Southern California (USC). (The justification for discussing this here, since it did not possess a general purpose digital computer, is that it is an instance of the concept which distinguishes this era from the previous one: the generation of the stepping cycle by a central logical unit not integral to the mechanics of the legs.)

THE PHONY PONY

Around 1965-66 R. McGhee, then at the USC, saw the General Electric Walking Truck and decided to demonstrate that such a machine could be controlled using a stepping cycle generated by a computer or electronic logic (the computers of the time were not very portable). At that time he was collaborating with Tomavic on the theory of finite-state control (Tomavic and McGhee 1966) and the machine which was built used this method. Computer control was seen as the way forward, beyond the dead end which both mechanically linked walking and pure teleoperator control seemed to have reached.

The Phony Pony was built by McGhee and A. A. Frank in 1966 (McGhee 1976) and was sometimes referred to as the Californian Horse. It was a quadruped weighing 50kg, with two degrees of freedom per leg (see Figure 1.8). Each joint was driven by a series-wound electric drill motor through a reduction gearbox and could be in one of three states: forward motion, backward motion, or locked. Limit sensors were provided for both extreme positions of each joint.

Each leg was controlled by an electronic sequencer made of flip-flops,

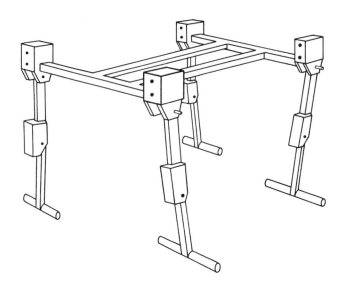

Figure 1.8 *The Phony Pony. A hip and a knee joint are visible;*
not shown is a passive load equalizing suspension mechanism: each hip
can move vertically, the two legs in each end pair
being connected by a spring.

which implemented a state machine with six or seven states. As the controller cycled through these states it carried out the leg movements making up the stages of a particular gait, either the quadruped crawl or the diagonal gait (trot). In the second case lateral stability depended on very wide feet (McGhee 1976). The four leg controllers were synchronized by being made to wait for interlock signals from the other legs. The state diagram for the diagonal gait is shown in Figure 1.9.

The Phony Pony's main importance is that its success led McGhee to go on to build more sophisticated walking machines. Frank also continued a separate line of legged locomotion research. McGhee's work in particular, through extensive international contact, has triggered off or revitalized many projects in several countries.

At about the same time (1966) a second computer-controlled machine was designed by R. J. Hoch and J. R. Kosook at Battelle-Northwest Laboratories. It used hydraulic actuation and had transducers for leg stress, body attitude, velocity and acceleration (Park and Fegley 1973). This project appears not to have been developed far or to have had much impact on subsequent research.

From 1969 onwards many research projects were started. McGhee and Frank took their interests in legged locomotion to Ohio State University and the University of Wisconsin, respectively. Vukobratovic

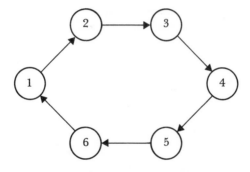

state number	hip state	knee state
1	locked	locked
2	rearward	locked
3	locked	locked
4	forward	rearward
5	forward	locked
6	locked	forward

Figure 1.9 *State diagram for diagonal gait (trot).*

and his associates in Belgrade worked on locomotion in bipeds; several Russian groups began work, mainly on hexapods; and other projects began in Italy, France and Japan. Some existing programmes such as that of Witt at Oxford continued into this period. Some of the more significant projects will now be described.

THE OHIO STATE UNIVERSITY (OSU) HEXAPOD

This machine first walked in 1977 (McGhee 1977) and is still in use. It is 1.3m long, 1.4m wide and weighs about 100kg (see Figure 1.10). Each of the six legs has three joints, two at the hip and one at the knee. Each joint is powered by a series-wound electric drill motor with internal gear reduction, driving the joint through extra reduction gearing. This includes a worm stage, so the joint cannot be back-driven. The maximum joint torque is about 300Nm (200ft lb).

The motors are powered from a 115V 60Hz supply using half-wave triac phase control. The triac trigger signals are produced by a PDP-11 computer (see Chapter 5) connected to the hexapod by a set of cables which also carry the signals from the sensors of joint angle, motor speed, leg force and body attitude.

The vehicle's speed is typically a few inches per second. It has been used for a great variety of experiments, such as: walking with different gaits on a level surface, climbing up shallow stairs and over obstacles, turning, active compliance, the use of sensors, and control by a multi-processor. It has also been a focus for research on methods of analysis and control. (Many of these investigations are described in later chapters.)

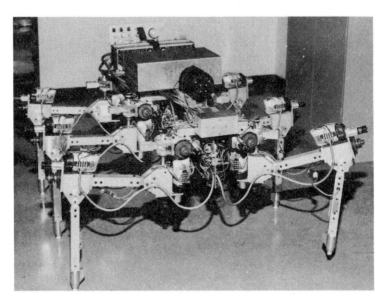

Figure 1.10 *The Ohio State University hexapod; the electric drill motor driving each joint can be clearly seen (courtesy Professor C. A. Klein).*

WORK AT THE UNIVERSITY OF WISCONSIN

In the late 1960s and the 1970s some work on exoskeletons was done (Grundman and Seireg 1977). A series of hydraulic exoskeletons for paraplegics was built, some of which were programmed for standing up, sitting down and stair climbing. A pneumatic three-legged robot was also built.

WALKING MACHINES IN THE USSR

The Russians began work on walking machines in the early 1970s. A hexapod with its legs disposed radially about a central vertical axis was built by Ignatiev in Leningrad (Vukobratovic 1973) but most of the work has been done in Moscow, at various related institutes. From 1971 onwards a series of papers on legged locomotion was published and by the mid-1970s two or three electric hexapods were in existence (Vukobratovic 1973; Okhotsimski 1979; Platonov 1979). Those described by Okhotsimski and Platonov are rather similar to the OSU hexapod; both were equipped with television cameras or rangefinders.

Perhaps the most significant contribution at this time was that of Bessonov and Umnov (1973, 1977) on defining optimal gaits, and that of Kugushev and Jaroshevskii (1975) on free gaits. Other theoretical work was that of Stepanenko on the dynamics of linkages (Stepanenko and Vukobratovic 1976). Recent work appears to have been mainly theoretical, on terrain adaptation (Bessonov and Umnov 1983; Devjanin *et al.* 1983).

BIPED RESEARCH IN YUGOSLAVIA

As mentioned earlier, Tomovic made one of the earliest theoretical studies of wheel-less locomotion (Tomovic 1961). From about 1969 onwards Vukobratovic and his colleagues at the Mikhail Pupin Institute, Belgrade have been active in the design and theory of bipeds and exoskeletal walking aids, starting with passive aids and going on to the study of autonomous locomotion. Vukobratovic introduced the concept of zero moment point control (see Chapter 4). In 1971-72 a pneumatically powered automatically walking exoskeleton was built (Vukobratovic *et al.* 1974) and tested with healthy and paraplegic users. It walked satisfactorily, although with the user providing additional stabilization with a pair of walking sticks.

RESEARCH IN WESTERN EUROPE

Walking machines have been built or proposed in several countries, but no very sustained effort has yet been made. The work of Hutchinson and of Witt in Britain in the era before computers has already been described; since then little has been done. M. W. Thring has made a number of experimental studies of walking (Thring 1983). Some of the machines built used spoked wheels or tracks rather than independent legs; he also experimented with purely mechanical legged vehicles and with exoskeletons. Computer control was not used.

Perhaps the most well-known project was that of Petternella of the University of Rome who, during the period from 1969 to 1974, was active in the walking machine field. He built an electric hexapod (Petternella and Salinari 1973; Vukobratovic 1973). In shape it resembled a table (as which, at the end of the project, it ended up). Its legs were unusual in employing telescopic joints for the vertical motion; they were pivoted at the hip to provide the propulsion stroke.

Some work on legged robots has been done in France; a current project is described in Chapter 6.

THE KYUSHU INSTITUTE OF TECHNOLOGY QUADRUPED

A pneumatic quadruped was built here in the mid-1970s (Taguchi *et al.* 1977). It overcame the difficulty of proportional control of pneumatics by using a chain of three actuators, regarded as binary devices whose strokes could be combined mechanically. As the speed of each cylinder was fixed by an orifice it was possible to select the time taken for a stroke by operating the three actuators in the chain sequentially or simultaneously. This three to one ratio was used to implement the quadruped crawl gait in which the machine balances on three legs while swinging the fourth. A chain of three cylinders was used for both the vertical and the horizontal movements, connected to the foot by a pantograph. This approach may have some utility for future low-cost robots but is inflexible.

THE WASEDA UNIVERSITY BIPEDS

One of the most photographed walking robots was the Wabot (Kato *et al.* 1974) which in addition to its legs had many other anthropomorphic features such as arms, head-mounted visual sensors and voice communications (see Figure 1.11). It first walked in 1973 and was statically stable at all times, relying on keeping its centre of mass above one of its large feet (or between them during the transition from one foot to the other). It was hydraulically powered.

Since then Kato and associates have built a series of bipeds, generally unencumbered by android paraphernalia, which have demonstrated a progression towards more dynamic forms of walking. These machines have generally been hydraulic, with a cylinder and servo valve for each joint. They have usually had about five joints in each leg. These bipeds are currently able to walk with a gait in which statically stable states alternate with unstable ones (see Chapter 4). A typical step length and cycle time are 15cm and 15s, respectively (Ogo *et al.* 1980). This research is still continuing.

Alternative Mechanisms of Locomotion

This chapter is primarily a history of locomotion with legs, but it would be incomplete without some mention of certain alternative locomotion mechanisms which have been tried for vehicles. The use of legs is only one of several approaches to achieving high mobility, and may not always be the best. Some alternative mechanisms for vehicles will now be considered.

VARIABLE GEOMETRY WHEELED AND TRACKED VEHICLES

Most mobile robots use wheels or tracks. Wheels have the advantages of engineering simplicity, low friction when on a smooth surface and,

Figure 1.11 *The Wabot (reproduced with permission from*
Robots and Telechirs: Manipulators with Memory; Remote Manipulators,
Machine Limbs for the Handicapped *by M. W. Thring published by*
Ellis Horwood, Chichester, England, 1983).

sometimes, of integral springiness. Tracks are a way of extending the use of wheels to soft and rough ground by laying down a track for wheels to run on.

Not surprisingly, given the enormous success of wheels and tracks, many attempts have been made to retain their advantages while overcoming their weaknesses. The main weakness is their poor performance when faced with a vertical step or a discontinuous surface. Of course, the problem can be solved by making the wheel large compared with the step or gap, but there may be factors limiting wheel size.

The simplest approach to improving a wheel or track is to give it some of the characteristics of a leg by attaching leg-like spokes to the rim. Many machines have been designed in this way; two of the simplest basic geometries are shown in Figure 1.12. These methods work, after a fashion, but give a lurching, unstable motion on stairs, particularly if required to turn at the same time.

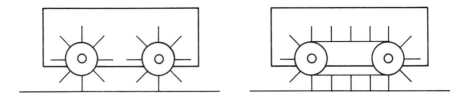

Figure 1.12 *Wheels and tracks with external spokes.*

An alternative approach is to introduce extra joints so that when one wheel comes to a discontinuity another wheel can be placed beyond the obstacle while the first is lifted over it. One way of doing this which can sometimes work well on stairs, and is sometimes used in hand-carts or sack trolleys, is shown in Figure 1.13. The wheels are in triplets, each triplet being free to rotate about a central axle. The vehicle is shown encountering a stair, with the front triplet rotating to bring a new wheel on to the first step. Methods of driving the individual wheels and the common axle vary.

Tracked vehicles have also been designed with extra joints (Trautwein 1973; Kohler *et al.* 1976; Iwamoto *et al.* 1983). The purposes of this are several: it can allow a sudden change from a flat surface to a very steep one; the body can be kept level on steep slopes; a transition is possible between a compact posture and a long but stable one; mobility on soft soil may be enhanced; and the ability to cross crevasses may be improved. Some examples are shown in Figure 1.14.

WHEEL-LEG HYBRIDS

Some machines may be regarded as variable geometry wheeled vehicles

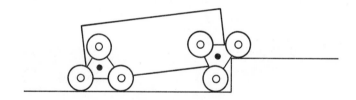

Figure 1.13 *Triangular clusters of wheels for stair climbing.*

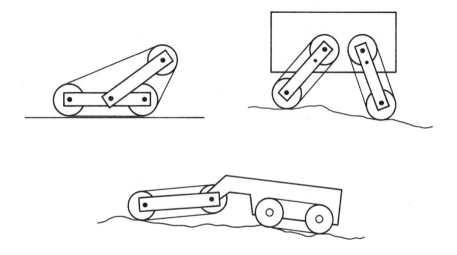

Figure 1.14 *Variable geometry tracked vehicles.*

or alternatively as walking machines with wheeled feet. An example (Ichikawa 1982, 1983; Ozaki *et al.* 1983) is a remote controlled robot for maintenance in nuclear power stations, designed by Hitachi (see Figure 1.15). It has five vertically telescoping legs each having a steerable, powered wheel on the end. On a smooth surface it simply rolls along on the wheels. To climb stairs it successively extends and retracts the legs to match the profile of the stair while rolling along one stairtread at a time. It can also roll up a ramp of up to 15°.

For inspecting the inside of small sewers and other pipes it is common to pass television cameras and other instruments through by pushing or pulling with a cable, but some work has been done on selfmoving instruments. The obvious feature dominating locomotion in pipes is that by pressing against opposite walls it is possible to generate a high force for providing traction. The vehicle may have tracks which press against the walls, or it may propel itself like an earthworm by

Figure 1.15 *The Hitachi wheel-leg hybrid robot (courtesy N. Ozaki, Hitachi Ltd).*
The leg vertical drive uses a leadscrew; the wheel on each foot can slide
a short distance as well as swivelling — this is necessary for climbing a step
because during the period when one leg is being lifted over the step
it is possible for another leg to collide with the step, preventing the machine
rolling any further. The wheel slide gives just enough extra movement
to get the raised wheel past the edge of the step.

alternately expanding a segment radially to jam it against the walls and contracting it and moving it forward while the vehicle presses against the wall with another segment.

A series of machines for pipes is described by Vertut *et al.* (1980), which use wheels to propel the machine along the inside or outside of the pipe and arm or leg-like levers to press against the inside or hold on to the outside of the pipe. Similar machines have been designed by Kemmochi and Kazuoka (1983) who also describe an octopod with electromagnet feet for walking on steel pressure vessels. This is a genuine, if specialized, walking machine. It uses the same way of steering as the Komatsu underwater robot (see Chapter 4).

ACTIVE CORD MECHANISMS

The most radical departure from previous methods of mechanical locomotion has been that of Hirose and Umetani at the Tokyo Institute of Technology. The potential utility of flexible structures like the snake, the elephant's trunk and the tentacle of the octopus has long been recognized, but these rely on large numbers of actuators, which if reproduced with current engineering methods, would be heavy and bulky. Hirose and Umetani have found several ways in which potentially useful flexible active structures (which they term 'active cord mechanisms') can be built. They have designed flexible-fingered grippers on this principle, and a series of experimental locomotion machines resembling a snake or centipede. An early one of these (Hirose and Umetani 1977) consisted of a train of short segments connected by hinge or ball joints, with a single azimuth actuator at each joint. (The machine was constrained to travel on a more or less plane surface.) It could travel between obstacles and through a labyrinth by thrusting backwards against the walls with waves of the body in the same way as a snake.

A more recent active cord mechanism can control its shape in three dimensions: for example, it can rear up while stabilized by a flat loop of the back part of its body (Hirose *et al.* 1983). It uses axial joints alternating with oblique-axis joints, as shown in Figure 1.16.

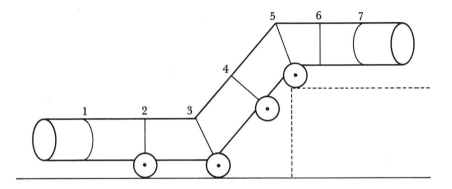

Figure 1.16 *Active cord mechanism: the machine is shown resting on its back three segments and rearing up to climb a step. The axial joints (2, 4 and 6) can be rotated so that the oblique joints lie in any plane: in this example joints 3 and 5 are in the plane of the paper and joints 1 and 7 are in the plane perpendicular to it.*

The designers suggest that for the relatively small number of segments which is practicable, the snake mechanism of generating thrust is unsatisfactory and it is better to provide a separate method, such as a pair of powered wheels at each joint.

Chapter 2

General principles of legged locomotion

The Theory of Land Locomotion in General

The types of movement in animals and plants may be categorized as growth, change of shape, and locomotion, which is the movement of the body as a whole. *Surface reaction locomotion* depends on the forces applied at a surface and includes locomotion in cracks, burrows and on flat, sloping and inverted surfaces. This requires the animal or machine (hereafter collectively referred to as a *mobile*) to exert, against the surface, a force having a component in the direction of motion (but opposite in sign, so that the propulsive force is a backward thrust). In general, there are four kinds of force involved (including propulsive thrust):

1) propulsion: thrust against the substrate in the direction of motion,
2) adhesion to allow anchorage for propulsive thrust,
3) adhesion to resist non-propulsive forces tending to detach the mobile from the substrate,
4) support to resist any forces tending to press the mobile against the substrate.

Note that this definition is intended mainly for surfaces of gentle inclination and, although consistent, is unnatural for inverted surfaces. In the case of a fly on a ceiling, for example, it could be said that it is supported by its legs, which grip the ceiling, but in terms of this definition the legs provide not support but adhesion.

Not all these forces need be present at once. For example, in running or hopping their occurrence alternates with periods of ballistic flight. It must also be noted that, in many cases, locomotion involves both surface reaction and fluid mechanical forces. A good example is the movement of arthropods on the sea bed.

Chapter 2 is confined to the locomotion of legged mobiles on open land. For some mechanisms of locomotion in other circumstances see Wells (1968). The image most readily generated by the phrase 'land locomotion' is of people, vehicles and domestic animals such as dogs and horses. These 'typical' examples belong, however, to a special environment: the solid surface of a planet where the acceleration due to gravity is $10m/s^2$, the viscosity of the atmosphere is very low and its density is $1kg/m^3$; and the dimensions of the mobiles themselves are a few metres or decimetres. This combination has several consequences: inertial and gravitational forces are of the same order, as are the forces required for support, adhesion and propulsion; and buoyant support by the atmosphere is negligible, as are its aerodynamic lift and drag except at very high speeds.

A change in any of these conditions alters the whole problem. An illustration is the case of gravity. Gravity is the engineer's curse for most purposes, being the principal force which many structures have to resist. However, for many mobiles gravity provides the main force of adhesion, via friction between the foot (or wheel) and the surface. In low gravity, fast agile locomotion is impossible, except when adhesion can be produced inertially, as might happen in a tight turn, or by non-frictional means such as jamming a foot against a protuberance.

To revert to the standard case of locomotion on the surface of the Earth, it is useful to ask why the only practical mechanisms are legs, wheels and tracks, and to compare their merits.

Locomotion requires a contact between the mobile and the ground which resists forces normal to the surface, thus providing support and adhesion. This contact must also be powerable so that a propulsion force can be exerted across it.

The most basic mechanisms fitting this requirement are the slide, the lever and the wheel (see Figure 2.1). The slide has been used for walking machines (see Chapter 6) but for the present purpose may be regarded as an infinitely long lever. The lever is the basis of the leg. A leg needs more than one lever because it allows continuous motion for only a short distance before it must be raised and brought forward, which can be done only with at least one more joint. In contrast, the wheel can run over an infinite distance. A caterpillar track is merely a way of providing a portable load-spreading track under a wheel. On unbroken ground at least, the difference between tracks and wheels is largely practical rather than theoretical.

The theory of wheels will now be summarized, before locomotion using levers is discussed.

The Soil Mechanics of Wheels

For a wheel, various theoretical models have been formulated which allow sinkage, motion resistance and maximum tractive force to be calculated. This has been done for a variety of kinds of wheel and types

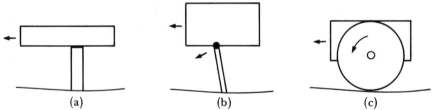

Figure 2.1 *Mechanisms allowing propulsion, support and adhesion*
(but not necessarily for an unlimited distance)
(a) slide; (b) lever; (c) wheel.

of soil. These models are discussed in Bekker (1956, 1969). To illustrate this method formulae will be given for a rigid wheel rolling over soft soil.

We shall assume that a rigid wheel rolls over the soil without slipping and sinks down by only a small fraction of its diameter. In this case the only resistance to motion is generated by the absorption of power in compressing the soil under the wheel, as shown in Figure 2.2.

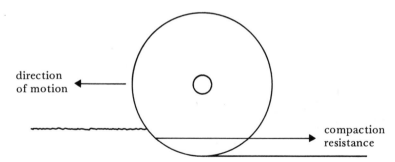

Figure 2.2 *Compaction resistance of a wheel.*

The backward, horizontal component of the resulting force is known as the compaction resistance. We shall ignore another source of resistance, the bulldozing resistance, which occurs when the wheel, having sunk until it is supported by the soil under it, moves through a viscous medium such as mud and has to push it aside.

To calculate the compaction resistance we need to know how the wheel sinks. The degree of sinkage is determined by the properties of the soil. Soil is generally regarded as having *cohesive* and *frictional* properties. Dry sand has mainly frictional properties, whereas clay supports large cohesive forces. Unlike cohesion, friction depends on load. These properties are expressed by Coulomb's equation:

$$\tau = c + p \tan \phi \tag{2.1}$$

where τ is the shear stress,

c is the coefficient of cohesion,

p is the ground pressure on the loaded area,

ϕ is a parameter called the angle of soil friction.

It has been found that the depth of sinkage of soil under a load can often be estimated by an equation of the form:

$$p = kz^n \qquad (2.2)$$

where p is the pressure under the wheel or foot,

k is a constant, the modulus of deformation,

z is the depth of sinkage,

n is an exponent which has been given the value 0, ½ or 1 by various workers.

The work per unit area is:

$$L = \int_0^z p\,dz \quad \text{or} \quad \frac{kz^{n+1}}{n+1} \qquad (2.3)$$

Equation (2.2) can be modified to take account of cohesion and friction separately:

$$p = (k_c/b + k_\phi)z^n \qquad (2.4)$$

where k_c and k_ϕ are the cohesive and frictional moduli of deformation and b is the smaller dimension of the loading area.

With these equations k can be determined experimentally. It is then possible to calculate the compaction resistance of a rolling wheel. Various formulae have been proposed. An example is:

$$R = \frac{W^2}{kbD} \qquad (2.5)$$

where D is the wheel's diameter and W is the load on it. This is a special case, for n = 0, of a general equation:

$$R = \frac{(3W)^{\frac{2n+2}{2n+1}}}{(3-n)^{\frac{2n+2}{2n+1}} (n+1)(k_c + bk_\phi)^{\frac{1}{2n+1}} D^{\frac{n+1}{2n+1}}} \qquad (2.6)$$

Sinkage can be calculated using the same model. A typical equation is:

$$z = \frac{3W}{(3-n)(k_c + bk_\phi)/D}^{\frac{2}{2n+1}} \qquad (2.7)$$

and the drawbar pull (DP) generated by a powered wheel can also be calculated. It is the difference between the soil thrust H and the motion resistance R:

$$DP = H - R \qquad (2.8)$$

and the limiting soil thrust depends on the soil properties described in Coulcomb's equation:

$$H \approx Ac + W \tan\phi \qquad (2.9)$$

where A is the area of contact.

This analysis is relevant to legs inasmuch as sinkage can be calculated in the same way, and soil compaction is a mechanism of thrust transmission.

Locomotion Using Levers

As described earlier, the only significant alternative to wheels is levers. Legs are not the only possible instrument of levered locomotion. In some animals the whole or part of the body is effectively shortened, lengthened and pivoted to produce levered propulsion. Certain aspects of levered locomotion, such as the interaction with the soil, apply regardless of what part of the system constitutes the set of levers, whereas others such as the loss of power in oscillating limb segments apply particularly to legs. This book concentrates on legs.

The fundamental problem with a leg is its limited step, so that it must repeatedly be lifted and swung forward for another stroke. Therefore, legged locomotion inevitably has a cyclic character in which each leg alternates between a support and propulsion stroke and a retraction or return stroke. (Note that in the zoological literature the term 'retraction of a limb' is related to the action of the main muscles used, so that the 'retraction' stroke may be the power stroke rather than the return one.)

A second implication is that during a leg's return stroke support and propulsion must be provided by some other means. Three such means have been found practical:

1) The mobile rests on its body between power strokes. This solution is adopted by very large walking excavators, where mechanical simplicity is more important than speed.
2) Several legs are used whose stepping cycles overlap so that there is always a source of support.

3) The mobile flies ballistically (possibly with some fluid dynamic support or stabilization) between propulsion strokes. This is observed in kangaroos: both legs can be regarded as a single one while hopping.

Two of the fundamental characteristics of legged locomotion will now be described. These are *soil mechanics* and *energetics.*

Soil Mechanics and the Differences between Feet and Wheels

Little work has been done on the soil mechanics of legged locomotion, certainly when compared with that on the adaptation to uneven, but hard, ground. Yet it may be that methods satisfactory on hard ground are ineffective on soft soil: consider the case of an antelope in a swamp. One of the few studies to test a walking machine on a variety of soils was the Iron Mule Train project described in Chapter 1.

Some of the main questions that can be answered using soil mechanics are:

1) Under what conditions is forward motion possible?
2) How does the foot move during the propulsion stroke?
3) Through what mechanisms is power dissipated at the foot-soil interface?
4) What are the mechanisms of ground damage?
5) How are balance, terrain adaptability and other control issues affected?

To answer such questions there are three possible approaches, which are better combined than used separately:

1) mathematical modelling: the foot is modelled as a suitable shape such as a plate or hemisphere and the deformation of the ground, considered as an elastic-plastic material, is calculated for various loadings and paths of the modelled foot;
2) physical models of the foot, loaded appropriately and instrumented, are pressed into or dragged over different types of soil;
3) tests on a complete vehicle.

The rest of this section shows how the interaction of foot and ground can be modelled, and compares feet and wheels.

The physical mechanisms underlying the aspects of performance previously listed are friction and soil deformation, as in the case of wheels. (This assumes a rigid foot.) It is of interest to compare the behaviour of legs and wheels, but there is no good way of doing this. One problem is that an infinite variety of wheel or leg arrangements is possible; for example, the number of wheels, their diameter, width, elasticity, tread, loading and suspension can be varied. It is obvious that on a smooth hard surface a vehicle with rigid wheels needs little power, but on soft or layered soils a comparison is much more difficult. Intuition is a poor guide. For example, a vehicle with wide tyres, which seems the natural choice for off-road travel, may have a higher rolling resistance and bog down more easily than one with narrow wheels. This happens if there is a shallow layer of soft soil on top of a hard layer. Narrow wheels cut more easily through the soft layer to roll firmly on the hard one, with reduced bulldozing resistance.

Figure 2.3 shows a simplified model of how wheels and feet may be regarded as special cases of an object moving over or through soil and being resisted by the forces of friction and compaction.

On hard ground there is no difference in principle between a foot and a wheel. The only issue is whether the foot or wheel slips or not, and the only parameter needed is the coefficient of friction (which is not to say that friction is a simple matter). On soft ground an unpowered foot will sink in, dissipating energy, but since it only does so in one place instead of having to be pushed along creating a rut, it should on average be more efficient than a wheel. A powered foot will usually travel further through the soil than an unpowered one, and with a horizontal component. It will dissipate power through soil compaction and friction on the bottom of the foot. But again, it does not have to create a continuous rut.

In many cases, on slightly compressible soil, the behaviour of a powered foot will resemble that shown in Figure 2.3h very briefly and then that shown in Figure 2.3f for the rest of the power stroke. This shows that the model of a foot's behaviour may include several phases such as initial penetration, the power stroke and withdrawal. This division would be particularly appropriate for a vehicle with small feet which would sink in quickly to a particular depth, then perhaps no further during the rest of the stroke.

Another difference between legged and wheeled locomotion can be seen by comparing Figure 2.3d and h. The thrust (drawbar pull) of a wheel is given by the difference between soil thrust and compaction resistance and may easily fall to zero if the ground friction is small, as in snow. But a foot is not limited in this way; it uses compaction resistance as part of its source of thrust and can in some conditions derive adequate propulsion from a completely friction-free surface. This is its most fundamental advantage over the wheel.

The Energetics of Legged Locomotion

The power used during locomotion on land is highly variable, the main sources of variation being slope, accelerations and environmental

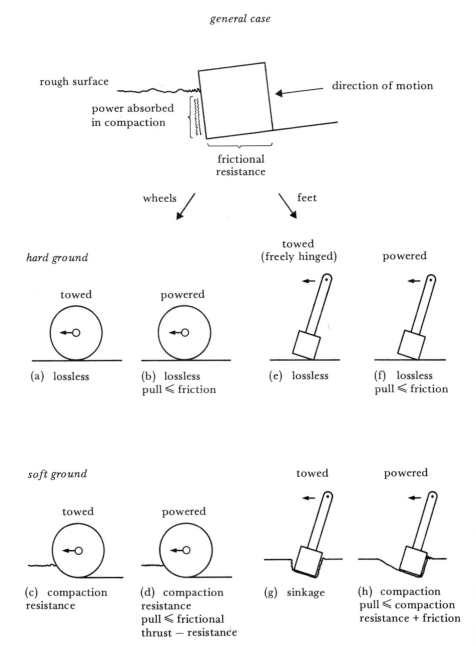

Figure 2.3 *The soil mechanics of feet compared with wheels.*
The general case (top) of an object being forced into a yielding material
can be divided into the cases of wheels (lower left) and feet (lower right).
For both wheels and feet the motion can be analysed on a hard or soft surface,
and for a propelling member or a passively supporting one,
giving four possible situations in each case.

resistance. The drive train of all vehicles must cope with these variations and this greatly complicates design.

Legged vehicles face the added difficulty that because the mechanical elements oscillate there must be continual (and not always regular) fluctuations in their kinetic energy, and it is hard to avoid losing part of it on each cycle.

Similar considerations apply to animals. The energetics of their locomotion has been extensively studied by, for example, R. McN. Alexander (Alexander 1980, 1982, 1984). The dynamics of human locomotion is reviewed in McMahon (1984). Bekker (1956) gives simple mechanical models of various forms of animal locomotion and uses these to compare their performance with that of existing vehicles. A simplified version of this comparison is shown in Figure 2.4. It includes the relationship proposed by Gabrielli and von Karman in 1950 which was supposed to be an empirical limit applying to land, sea and air transport.

A useful measure for comparing vehicles or animals is the *specific resistance*. In the terms of Figure 2.4, this may be defined as:

$$\epsilon = \frac{\text{unit power}}{\text{speed}} \quad \text{or} \quad \frac{\text{power}}{\text{weight} \times \text{speed}} \quad (2.10)$$

or it may be defined (Hirose 1984) as:

$$\epsilon = \frac{E}{W \times L} \quad (2.11)$$

where W is weight and E is the energy consumed in travelling a distance L. Hirose distinguishes four variants of ϵ, depending on whether W is the total weight or payload and whether E is the total energy or power consumed by the mobile or only the fraction actually used for locomotion. He suggests that the most appropriate measure for comparing walking machines is:

$$\epsilon^* = \frac{\text{net or propulsion energy}}{(\text{payload weight}) \times (\text{distance})} \quad (2.12)$$

Before discussing mechanical efficiency, it should be noted that it is not always very important. The penalties associated with poor efficiency are:

1) fuel costs,
2) battery cost if appropriate,
3) weight penalty due to increased need for fuel or batteries,
4) weight penalty due to increased engine size,
5) all these are increased yet again because of the force inefficiency of actuators; because of friction they have to be bigger than they otherwise would.

However, for many robot applications some of these factors are

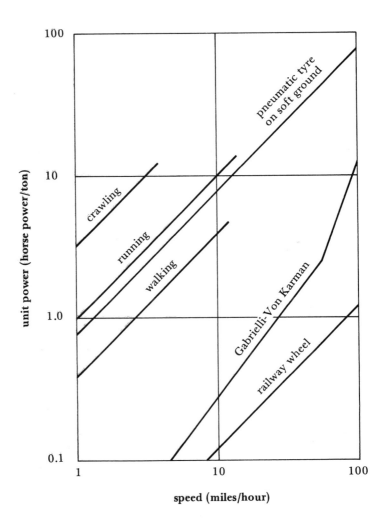

Figure 2.4 *Unit power as a function of speed.*

unimportant. For example, if power is supplied through a cable, there is no fuel or battery weight problem. If the cost of fuel is small compared with the benefits of using the robot, power consumption does not matter. In the near future, it is unlikely that legged robots will compete directly with existing vehicles, but will tend to be used for purposes where the unique qualities of legs are the overriding consideration. Nevertheless, at present the problems of obtaining an adequate power-to-weight ratio for a legged robot are so severe that in most cases the designer cannot afford to neglect efficiency.

THE ORIGINS OF ENERGY LOSSES

The main ways in which energy is dissipated in legged locomotion are as follows:

1) Loss of the kinetic energy which must be imparted to the whole machine to make it leap. When the machine lands it may come to a complete halt, or it may keep much of its forward speed, so the energy lost may be that associated with just the vertical component in kinetic and potential energy. The net power loss depends on the relative phase of the components, which in turn depends on the gait (McMahon 1984).

2) Loss of the kinetic energy which must be imparted to the limbs to make them oscillate.

3) Soil compaction, or slipping on the surface, and other forms of motion resistance, such as bulldozing or that due to pushing through vegetation.

4) Power wasted (in some cases) in supporting the body against gravity or other constant forces. This is more important for actuators such as muscles which dissipate power when loaded even at rest.

5) Geometric work: Figure 2.5, based on Waldron and Kinzel (1983), shows a mammalian-type leg at two successive positions during the stride. In Figure 2.5b, the hip actuator must do work against the direction of motion of the thigh, to brake it at the end of its stroke. This is called geometric work.

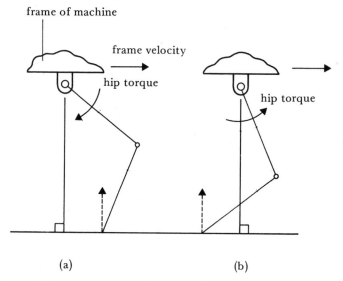

Figure 2.5 *Geometric work: direction of joint actuator torques; (a) early in the support phase; (b) late in the support phase; here the hip actuator has to do work against its direction of motion, wasting energy.*

6) Inefficiency of actuators and power source, which applies to all vehicles, legged or otherwise.
7) Opposing actuators: it may happen, particularly if there are many legs, that they do not all thrust in exactly the same direction or at the same speed. They will then oppose each other to some extent and waste power in the process. This problem does not appear to occur in most animals, which have excellent sensing of limb forces and control of them.
8) Any failure to achieve smooth motion of the body when walking on rough ground will waste energy as the body is continually made to go up and down.

FACTORS AFFECTING ENERGETIC EFFICIENCY AND METHODS OF IMPROVING IT

This section expands on certain items listed above, and then discusses ways of saving energy.

One of the main determinants of efficiency is the type of actuation used. If animals are included for comparison, the available methods are:

1) muscle,
2) mechanical transmission,
3) pneumatic,
4) hydraulic,
5) electric.

It may be noted that in some of these situations force or torque is generated at the joint itself, whereas in others it is transmitted from a central engine.

Mechanical transmission from a central motor has not been achieved in a way allowing any flexibility of control. It has the advantages of high efficiency and lightness, but it is not clear whether a regenerative mechanical drive would be practical. Presumably, energy would have to be stored in springs or flywheels. Also, unless there is a motor per joint, speed control of each joint individually would need efficient, continuously variable, drives.

Pneumatic actuation has been used, partly to take advantage of the elasticity of air for energy recovery, and partly because of its relatively economic and simple use. (It ceases, however, to be simple if any degree of positional or velocity servo control is required.) Unfortunately, most air compressors are big and inefficient.

Electric and hydraulic actuation can take many forms. Waldron and Kinzel (1983) list two main classes of hydraulic actuation and three categories of electric actuation, which vary greatly in efficiency. A servo valve controlled hydraulic system dissipates high levels of power in the valves and regeneration is not easily achieved. In a hydrostatic system there is a circuit, consisting of a variable displacement pump and an actuator, for each joint. This is, to simplify, efficient over a range of speeds, regenerative, but possibly heavy and slower in response than a valve system. The Carnegie-Mellon hexapod uses a compromise which seems effective; a few variable displacement pumps control speed and low-loss on-off valves select direction.

Electric actuators can use direct current (d.c.) servo motors (inefficient) or alternating current (a.c.) phase-controlled motors. In both cases there is usually a gear train which introduces friction and cannot be backdriven (and, therefore, does not allow regeneration). This is not necessarily true for all gearing mechanisms.

A second factor in determining the energetic efficiency of a legged vehicle is the geometry of the legs. An important example is the problem of geometric work described earlier. This can be avoided by decoupling the horizontal (propulsion) and vertical (lift) actions so that the propulsion stroke requires movement by just one actuator, and also does not induce any vertical component of body movement. On a flat surface such decoupling implies that no power is needed for the propulsion stroke apart from overcoming inertial and dissipative forces. On a slope the propulsion stroke must do gravitational work, but at least this is at a steady rate. Several methods of decoupling are described later.

A design issue which can perhaps be classified as geometric is the number of legs. Hirose and Umetani (1980) calculated that under some circumstances power consumption is inversely proportional to the number of legs, but this conclusion seems unlikely to be universal. It is always difficult to dissociate the number of legs from other aspects of the design. The same sort of uncertainty attends attempts to correlate power consumption with gait (that is, when there are alternative gaits for a given speed).

Turning now to energy-saving methods, there are three basic principles:

1) minimize dissipative losses,
2) minimize the diversion of energy into unproductive forms such as the kinetic energy of limbs,
3) recover energy whenever possible.

Dissipative losses have been discussed elsewhere; they arise mainly from the inefficiency of power transmission and from environmental resistance. This section discusses ways of minimizing non-dissipative losses. (Of course, in the long run all energy loss is dissipative, but the distinction being made here is between losses in which energy is dissipated directly, through friction and so on, and those in which it is first transferred to some reservoir from which it could in principle be recovered.)

The most obvious way in which energy is diverted into an unprofitable form is, as mentioned previously, the need to repeatedly accelerate and decelerate each leg. Many attempts have been made to minimize this by (a) reducing the mass of the leg; (b) devising linkages which allow the heavy actuators to be fixed to the body and (c) using leg cycles which minimize acceleration. Some of the techniques described later, which are of more general use, can also be applied to leg energy conservation.

One of the most important energy-saving techniques is to encourage the transfer of useless energy to a reservoir from which it can be recovered. The most obvious example is the interchange between kinetic and potential energy. The potential energy can be gravitational or elastic. An example is the Carnegie-Mellon hopper (see Chapters 4 and 6) in which at landing and take-off there is an interchange between the kinetic energy of vertical body motion and energy of compression of a pneumatic spring. Systems such as these are resonant, and power consumption is minimized by conforming to the natural frequency. This example is complicated by the fact that during flight it is a gravity/ kinetic resonant system, whereas during the stance phase it is a spring/ kinetic resonant system.

This machine also illustrates another aspect of energy conservation, that is if forward speed can be kept constant then no energy is wasted on accelerating or decelerating it.

Resonance can be applied to the whole machine, as in the example of the hopper, or to some part of it. For example, when man walks the non-supporting leg acts as a pendulum to some extent.

The exchange between kinetic and potential energy is not the only kind possible. Often the kinetic energy of a limb is exchanged with that of the body; such exchanges are used in an endless variety of ways by athletes.

Another design aspect relevant to energy conservation is the balancing of one limb movement against another, for example swinging one leg forward while another swings back. The net effect is to minimize the disturbance of the body. Even when this does not directly affect power consumption it may make control easier and give a smoother ride, which in turn may result in less power being used for correcting body speed or attitude. Some animals use the head and neck or the tail to balance the torques and thrusts produced by the legs. In the process, the head or tail becomes a temporary store of kinetic energy, thus combining the functions of balance and energy conservation.

Finally, a mechanical approach to energy conservation is to use regenerative drives. An example is the hydraulic leg actuator which, if there are intervals when it is allowed to be backdriven by other forces, temporarily acts as a pump, and returns some energy to the hydraulic system. This is in principle one way of recovering the kinetic energy of a moving leg when it must be stopped. In practice, it is very difficult to achieve significant savings in this way.

Other Properties of Legs

This section describes some miscellaneous aspects of legged locomotion. First, it may be noted that the mobility advantage of legs over wheels is less a consequence of better soil mechanics than of the flexibility in choice of foot placement and load distribution among the legs. A legged robot can choose where to put its feet; and if a foothold proves inadequate there is the chance of shifting the foot to a new one. This provides a means of utilizing isolated firm footholds and of avoiding getting bogged down. It may also be able to rest its weight on those feet which are securely placed. These mobility advantages apply only to machines which can sense the relevant effects such as sinkage, slipping and load distribution, and are flexible in their foot placement. It also helps if they can respond quickly to the threat of overturning or other sudden problems. Also, to make full use of flexible foot placement it is necessary to be able to detect good footholds in advance. None of these abilities are possessed by fixed cycle machines, such as the Iron Mule Train, which accounts for their lack of success. A related ability is the ability to step over or between obstacles.

Another property of legged locomotion is the ability to control body height and posture, to some extent independently of fluctuations in the ground height and of its average slope. This is discussed in Chapter 4.

Finally, a leg can be a multi-purpose instrument. Virtually all animals use some of their limbs both as legs and as manipulators or for other functions. Even the most stolidly quadrupedal herbivores will use their fore legs to paw the ground or grasp a mate. This has hardly been explored at all as an engineering concept, partly because of the difficulty of making machines walk at all, and partly because the standard engineering approach would be to fit a legged vehicle with separate arms if required (resulting, perhaps, in a machine resembling a centaur).

A complementary point is that parts of the animal or vehicle, such as arms, wings, tail or head and neck, can be used to assist locomotion by levering against the ground, grasping trees, acting as inertia paddles or moving the centre of gravity to improve balance. Such mechanisms are especially important if it is required to sit up, or recover from falling over, which seems a desirable attribute in a rough-terrain vehicle. They are discussed in the next section.

Mechanisms Other Than Legs

Until now the technique of mechanical walking has been in such an undeveloped state that it would have been premature to pay much attention to the design of complete systems of which the walking mechanism is merely a part. But now that the first practical legged vehicles are in sight, it is worth looking at what sorts of system are possible and what problems, in addition to the mechanics of walking, will have to be solved.

The idea that walking machines are a kind of *vehicle* leads to the expectation that they will resemble trucks or tanks with legs. Many will indeed take this form, but it does seem rather unimaginative. If a persuasive case can be made for the advantages of legs over wheels, it seems logical to see how much further the analogy with biology can be taken, and ask what other animal mechanisms could usefully be built into legged machines. Of course, extra animal-like joints are not unprecedented; the best known example is the anthropmorphic Wabot (Kato *et al.* 1974), and the idea of putting manipulators on a legged robot is old, but in general non-leg joints have rarely been used.

There are two distinct reasons for extra mechanisms. First, the machine may have to carry some mechanism unconnected with loco-motion, such as a crane or excavator arm. Second, locomotion itself may be helped by extra mechanisms for balancing, pushing against the ground and so on. Devices such as manipulators and excavators may occasionally assist with locomotion, and their design should take this into account. Conversely, the legs may be used for functions other than walking.

This is an appropriate point to digress slightly by recording a trend in the design of wheeled vehicles towards a more variable geometry. A minor example is the ability of some trucks to raise one set of wheels off the ground when not heavily loaded; more significant is the steady increase in the number and diversity of (mostly hydraulic) cranes and other lifting devices, and powered joints in the vehicle body. Extrapolating this trend, one can envisage vehicles ceasing to be rigid boxes and becoming many-jointed, almost flexible, structures which can fit themselves around a load or be equipped with a variety of manipulators. If this is true of wheeled vehicles it must surely happen to legged ones as well.

Aids to Locomotion

SPINE JOINTS

Chapter 3 describes cursorial adaptations in which spine flexure is used as a way of increasing stride length. A similar result could be achieved by a telescopic expansion and contraction of the body. Such mechanisms only work for certain types of gait. The method does not seem suitable for a hexapod.

BALANCING AIDS

Machines such as Ohio State University's ASV or Hirose's PVII can to some extent adapt to sloping ground by adjusting differential leg height or average leg angle, but this may have undesirable consequences such as limiting stride length. An alternative or supplementary method of balancing on slopes is to move the centre of mass of the machine by extending, swivelling or bending extra appendages. The problems of balance in the sagittal and frontal planes are similar, but an asymmetry is introduced by the fact that it is undesirable to rely on long members sticking out sideways as they would prevent passage through narrow gaps, whereas long extensions at the ends are less awkward. Some examples of possible balancing devices are shown in Figure 2.6. Figure 2.6a and c could be combined into a single mechanism in which a flexible appendage can bend both up and down and from side to side, moving the centre of mass both longitudinally and laterally (see Figure 2.7). This might be a good application for an active cord mechanism. Such balancing devices might also be useful in obstacle crossing, by allowing the machine to use a normally unstable gait (see Figure 2.8).

PUSHING AND PULLING AGAINST THE SURFACE

A flexible or jointed neck and tail could have additional uses. One class of uses involves reaction against the ground. A simple example is as an emergency device to prevent the robot falling over on slopes; if the machine is in danger of overbalancing it can brace its neck or tail against the ground while it shifts to a new foothold. A more dramatic possibility is their use for returning the robot to its feet should it fall over completely (see Figure 2.9).

A third example of ground reaction is pulling the vehicle up slopes and out of ditches. This, of course, requires that the flexible extension is capable of gripping a tree, say, and powerful enough to take the weight of the machine.

In passing, it is worth noting that animals use a variety of aids to walking, such as dragging part of the body along the ground (crocodiles on mud) or using one or two legs as passive stabilizing skids (some crabs).

Non-Locomotory Uses of Appendages

First, it may be remarked that the legs themselves can have functions other than walking, such as digging or manipulation. This has hardly ever been tried; one exception is the Odex I, whose legs have been used to lift objects (Russell 1983), although the absence of any gripping device must limit the effectiveness of a leg when used in this way.

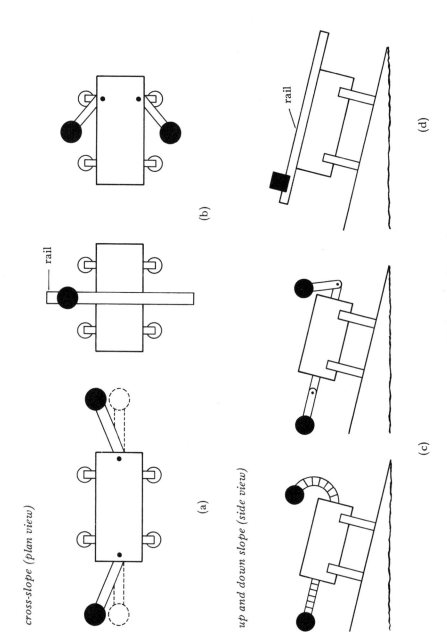

cross-slope (plan view)

(a)

(b)

up and down slope (side view)

(c)

(d)

rail

rail

Figure 2.6 *The use of balancing aids on slopes.*

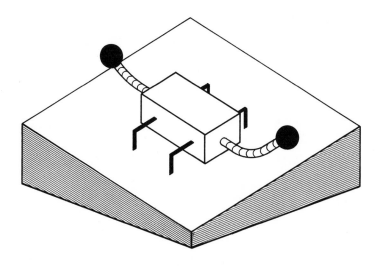

Figure 2.7 *The use of a flexible neck and tail
for both lateral and longitudinal balance.*

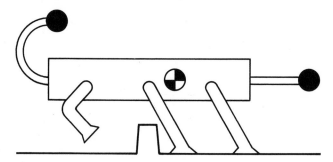

Figure 2.8 *The use of balancing aids in crossing an obstacle.
The head and tail masses move the centre of mass of the machine back
while the front end is unsupported; then during the second half of the manoeuvre
they swing forward to balance the unsupported rear of the machine.*

end view

Figure 2.9 *The use of neck and tail-like appendages
to restore an overturned robot to its feet.*

For them to be useful in such roles extra joints and a greater range of movement of the existing joints are probably needed.

A function which can be achieved either by collective action of the legs or by introducing extra joints is changing the shape of the machine. Being able to change the shape is useful for adapting to different tasks such as walking through a narrow passage, climbing a steep slope or lifting a heavy object. The best example of shape changing using the legs is again the Odex I, whose joints all have large angular ranges and whose leg arrangement lends itself to shape changes (see Figure 2.10).

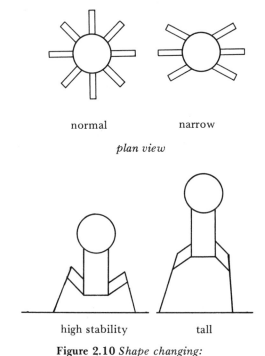

normal   narrow

plan view

high stability   tall

Figure 2.10 *Shape changing:*
(a) plan view of a hexapod in normal and narrow configuration;
(b) side view in low, squat and tall, small floor area configurations.

Shape changing by extra joints has not been used with walking machines but has been tried with tracked vehicles (see Chapter 1).

Another non-locomotory use of appendages employs extra mechanisms designed for some function such as digging or lifting. Such devices, in particular cranes and excavators, are of course familiar on wheeled vehicles. Some possible configurations are shown in Figure 2.11.

These appendages are all articulated and bear some resemblance to a tail, or head and neck. This suggests that with proper design they could also fulfil some of the roles, such as balancing, discussed in the previous section.

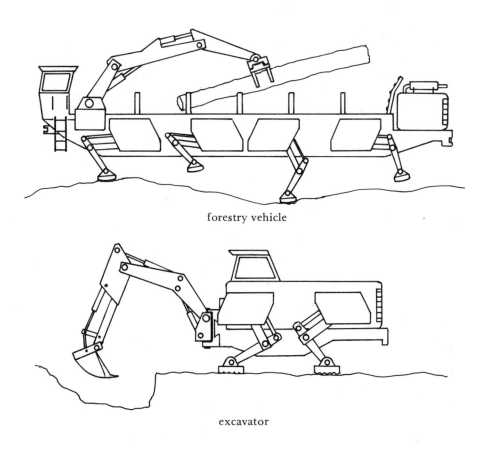

forestry vehicle

excavator

Figure 2.11 *Applications of appendages.*

Description and Classification of Gaits

A gait is a repetitive pattern of foot placements. It is usual to assume that each leg is sufficiently specified as a two-state device, the states being on the ground and off it. The legs on the ground are supporting and propelling the mobile, and those in the air are being retracted. The concept of gait assumes a regular progression forwards or backwards. Clearly, there are times when an animal stands still or shuffles in an irregular way, and the idea of gait has no utility at these times. In theory, smooth progression of the mobile is possible without a regular gait if there are enough legs; it would be possible for each leg to cycle almost independently of the others. This has some merit on uneven ground with few good stepping places. There is obviously a limit to the independence which can be permitted, or times will occur when there are not enough legs on the ground for stability. This is discussed in Chapter 4.

The reason for interest in gait is that it is important in the design of a legged vehicle to determine such things as whether there are families of equivalent gaits, or whether certain gaits are optimal in any way. There are three main approaches to the study of gait:

1) The geometry of straight-line gaits, for example the classification of gaits, and analysis of their stability.
2) The extension of this to the more complicated cases of curved paths and uneven ground. A practical walking machine must be able to choose where to put its feet under these circumstances.
3) Dynamics: for low speeds, the preceding two can be treated statically, but for balancing at higher speeds and for estimating power consumption the dynamic interaction of the body and the limb segments, which are repeatedly accelerated and decelerated, must be taken into account.

This section is confined to the first approach. It introduces some common ways of describing and classifying gaits.

A gait can be expressed as a function of time or distance. This is illustrated in Figure 2.12. Figure 2.12a shows a plan view of a walker's footfalls at four successive times, t_1, t_2, t_3 and t_4, as it walks with a 'diagonal gait' or 'trot' in which the left front and right back legs move as a pair and the right front and left back as a second pair. At times t_2 and t_4 the walker changes from one support pair to the other. Figure 2.12b shows the gait as a function of time. Each bar represents the time during which a foot is on the ground. This is called a *gait diagram.*

Other ways of expressing gaits have been used. A simple but restricted one is the support formula in which only the number of supporting feet is recorded, in sequence. The gait described above would be recorded as 4-2-4-2-4-2... Another way is the *footfall formula*, which is an abstracted form of Figure 2.12b, which for this example is shown in Figure 2.13. Although in common use for animals, they convey less information than a gait diagram. Gait diagrams and footfall formulae are sometimes expressed in a circular form as this accords well with the cyclic nature of gait.

In addition to these graphical representations, a quantitative description is needed. Therefore, the following terms need to be defined. A *stride* is a complete cycle of leg movements from a particular leg movement to the next occurrence of the same leg movement. (It is assumed that each foot is lowered only once during each stride.) The *stride frequency* f is the number of strides in unit time. The *stride length* λ is the distance travelled by the body during one stride. So the mean

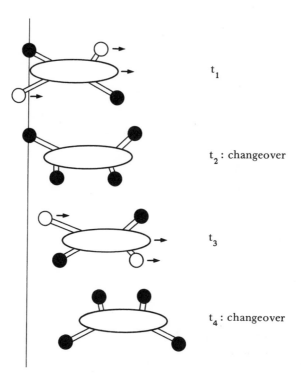

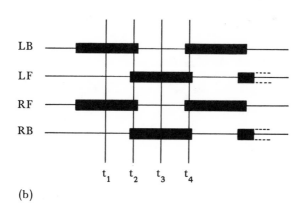

(a)   closed circles represent a foot on the ground

(b)

Figure 2.12 *Interpretation of a gait diagram.*

speed u is equal to λf. The *duty factor* β of a foot is the fraction of the cycle for which it is on the ground. It is often the same for all legs. There is a minimum value of β for statically stable walking for a given number of legs; this is 0.75 for four legs, 0.5 for six legs and 0.375 for eight legs. The following quantities also need to be defined:

T: cycle time, or period,
S: length of propulsion stroke of a leg,
t_{prop} : duration of the propulsion stroke,
t_{rec} : duration of the recovery or return stroke.

This notation is partly based on those of Alexander (1984) and Waldron *et al.* (1984).

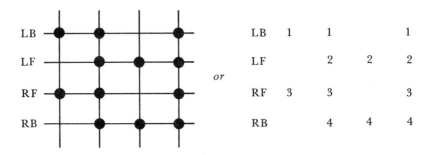

Figure 2.13 *Example of the footfall formula.*

Since a stroke of length S must take place in time t_{prop} :

$$u = \frac{S}{t_{prop}} \tag{2.13}$$

The times are related by the equations:

$$t_{prop} + t_{rec} = T \tag{2.14}$$

$$t_{prop} = \beta T \tag{2.15}$$

$$t_{rec} = T(1 - \beta) \tag{2.16}$$

These allow conclusions to be drawn. For example, Waldron *et al.* (1984) show that assuming a fixed value of t_{rec}, determined by the maximum leg speed, the following relationship holds between vehicle speed and recovery time:

$$u = \frac{S}{t_{rec}} \left(\frac{1-\beta}{\beta} \right) \tag{2.17}$$

Given the minimum values of β listed above this shows that the speed of a vehicle with four, six or eight legs is $S/3t_{rec}$, S/t_{rec} or $5S/3t_{rec}$, respectively, which allows the speed to be increased as the number of legs increases.

To continue with definitions, the *relative phase* ϕ_i of a leg i is the fraction of the cycle between the touch-down of a reference foot (usually the front left) and the touch-down of foot i. (Alexander defines the reference foot as that which is set down less than half a cycle before the other in the pair, so that the descriptions of mirror image gaits are similar.) A *symmetric* gait is one in which the left and right feet of a pair have equal duty factors, and relative phases differing by 0.5. An example of the use of these quantities in the description of a specific gait is given in Chapter 4.

An enormous number of gaits are theoretically possible, if any sequence of leg movements is allowed. McGhee (1968) introduced a notation, similar to the footfall formula, called the *event sequence*. The legs are numbered 1 to k. The event of placing leg i is denoted event i, while the lifting of leg i is denoted i + k. The gait in the previous example (see Figures 2.12 and 2.13) can be expressed as 24571368 (or 42753186 since there is no ordering within each pair). Gaits like this, in which some events occur simultaneously, are called *singular*, as opposed to totally ordered gaits. McGhee showed that the number of distinct totally event sequences for a k-legged mobile is:

$$N = (2K - 1)! \qquad (2.18)$$

which for a hexapod is 11! = 39,916,800. This number can be reduced by imposing various plausible constraints, and gaits can be classified and enumerated on this basis. In practice, only a few gaits are studied. There are two main sources of useful gaits: observations of animals and optimality calculations. This means that certain gaits can be shown to optimize some aspect of performance.

The most studied configurations are the quadruped and the hexapod. For bipeds the concept of gait at the level expressed in a gait diagram contributes only a little to the understanding of locomotion. Treatments of human locomotion are almost always in terms of dynamics — see, for example, McMahon (1984).

Quadruped gaits have been classified by Hildebrand (1976). He discusses the variables needed to characterize a gait and introduces the *gait graph*. This is a plot of a fraction of the stride by which fore footfall follows hind footfall against the duty factor, and is a useful way of classifying animal gaits.

The nomenclature of gait is somewhat variable (Hildebrand, 1976). For the present purpose, it is enough to make one or two distinctions.

A run is a gait with a duty factor of less than 0.5, so there are stages when all four feet are off the ground, whereas for a walk it is greater than 0.5. Alexander (1984) defines eight quadrupedal running gaits in terms of relative phase. However, some gaits, such as the trot, can fall into either category; looking at Figure 2.12b, which is the gait diagram of a walking trot, it can be seen that by shortening the bars slightly there would be times when all four feet were off the ground and the trot would be a run. The trot is of interest as it is one of the two simplest practicable gaits for a machine. Provided the problem of lateral stability can be overcome, it offers a straightforward way of continuously increasing speed and making the transition from walking to running.

The other important gait for quadrupedal machines is the quadruped crawl. (There seems to be no generally accepted definition of a crawl, but for a quadruped it could perhaps be defined as any gait in which only one leg is lifted at once. This implies a duty factor of at least 0.75.) The quadruped crawl is an example of a gait fulfilling an optimality criterion. The criterion it maximizes is called the *longitudinal stability margin*, which is a measure of the stability against falling over forwards or backwards. It is the minimum distance in the direction of motion of the vertical projection of the centre of gravity from the front or back boundary of the support pattern. It assumes that acceleration and deceleration are not significant and, therefore, applies only at low or constant speed. Maximization of this criterion leads to a unique optimum gait for quadrupeds at low speeds, as shown in Figure 2.14.

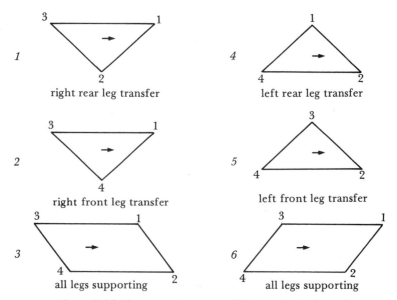

Figure 2.14 *Six successive stages of the quadruped crawl:*
(a) right rear leg transfer; (b) right front leg transfer; (c) all legs supporting;
(d) left rear leg transfer; (e) left front leg transfer; (f) all legs supporting.
The legs are numbered as follows: 1 left front; 2 right front; 3 left rear; 4 right rear.
The arrow shows the movement of the body during the stage.

This gait, called the quadruped crawl, is one of a class called *wave gaits*, characterized by a forward wave of stepping action on each side of the body, with a half-cycle phase shift between the two members of any right-left pair.

For hexapods, maximization of the longitudinal stability margin also leads to wave gaits. The generation of a hexapod wave gait is described in Chapter 4. In six-legged animals more legs are usually lifted from the ground as the speed increases. In insects we can see this progression from a one-leg-moving-at-a-time gait to an alternating triangle gait. (This is not true of all insects.) This progression is shown in Figure 3.11 in the form of a gait diagram. It is based on Wilson's (1966) model of locomotion, in which the different gaits are generated by varying degrees of overlap of the basic rear-to-front rhythm whose basic form is shown in Figure 2.15.

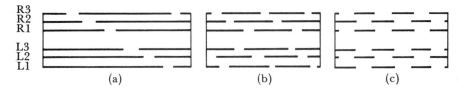

Figure 2.15 *Progressive stages of a wave gait: (a) one leg at a time (crawl); (b) an intermediate wave gait; (c) alternating triangle gait.*

The alternating triangle gait shown in Figure 2.15c is particularly important for walking robots, and accounts for the prevalence of six-legged machines. Six is the smallest number which always provides a tripod for support even when half the legs are raised. It therefore allows reasonably fast walking, while maintaining static stability at all times.

Examples of the gaits of six- and eight-legged animals can be found in *Locomotion and Energetics in Arthropods* (Herreid and Fourtner). Some of the literature (Pearson and Franklin 1984) discusses turning, starting and stopping as well as steady straight-line walking. It shows that, as might be expected, regular gaits are not always an adequate model of leg motion. Arthropods often change gait as they speed up, using many legs at low speeds and sometimes only two or four when running. One or two legs may be trailed along the ground as stabilizers. In an animal with many legs, such as a centipede, a collective style of control, in which waves of leg action sweep from one end of the animal to the other, is found.

Chapter 3
The mechanics of legged vehicles

Animal Structures and Their Applications to Robots

Throughout the short history of legged robots, animal locomotion has been a source of inspiration, and many features initially observed in animals have been built into robots. However, the locomotion of many animals remains infinitely more advanced and more elegant than that of present-day robots. It may seem that this will be so forever, but we can at least analyse the structure and behaviour of animals to see why they are so efficient, and what we can learn from them. It will then be possible to determine which aspects of animal locomotion can be imitated or improved and which aspects depend on biological properties that cannot be emulated.

Gait has already been discussed in Chapter 2; it is a concept derived from the study of animal locomotion. Chapter 3 concentrates on structure and geometry; structural adaptations are discussed first.

STRUCTURAL ADAPTATIONS

All animals are adapted on different levels. For example, at a major level the possession of legs is an adaptation to the need for transport on a solid surface; at a more detailed level the mammalian arrangement of tucking the legs under the body is an adaptation for energetic efficiency; and at a more detailed level still we observe that the proportions of a cheetah compared with those of a lion, for instance, represent an adaptation for speed. By studying animals we can discover some relationships between structure and performance. Some of these relationships cannot be applied to mechanical structures, but others can be used. This section presents a miscellany of adaptations which may be of use to the robot designer.

Mammals and Reptiles

The first subject to be discussed under this heading is the instructive difference in construction between mammals and reptiles. To over-simplify, mammals and reptiles have their legs arranged as shown in Figure 3.1.

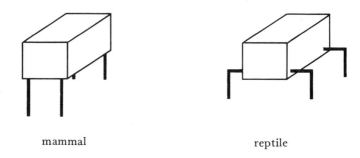

mammal   reptile

Figure 3.1 *Comparison of mammalian and reptilian leg arrangement.*

The reptilian stance provides a wide and, therefore, stable base and keeps the centre of gravity low. The body can easily be lowered slightly to rest on the ground. Such an arrangement is suited to slithering over mud banks and under logs. However, it has disadvantages:

1) When standing on diagonally opposite legs, there is a large twisting movement about the long axis of the body, which has to resist it; and the pelvic and pectoral girdles have to be rigid and box-like.

2) Energy must be expended at the hip to support the weight of the body even when standing still. (This ceases to be true for actuators that can sustain a static load with no power consumption.)

The mammal avoids these problems but requires more careful control of stability. Keeping the legs almost vertical minimizes the bending stress on the upper leg bone and the muscular power consumed in supporting the body.

Graviportal Adaptations

The mammal's basically vertical leg orientation may be regarded as a graviportal or weight-carrying adaptation. It is most important in heavy animals, and the departure from the vertical is, therefore, minimal in animals such as elephants. The requirement to keep the legs almost vertical most of the time imposes constraints on gait, as does the need to keep shock loads to a minimum because the bones of such animals are relatively highly stressed. Heavy animals are adapted in additional ways for weight carrying, but most of these are not relevant to an artificial structure.

Cursorial Adaptations

Cursorial animals are those which can move quickly, usually by running.

Animals that jump or hop are said to be saltatorial (Hildebrand 1974, from which must of this section is derived).

Speed is the product of stride length and rate of stride. The stride in this definition may be a leap or hop if the animal leaves the ground. Therefore, high speed can be achieved by a long stride or rapid striding, or both. The rate of stride is limited for any actuation method, including muscle, so adaptations for speed are directed mainly at increasing the length of stride. Not surprisingly, if an animal jumps, as does a kangaroo or a frog, its stride length can be several times its body length; but this is also true of animals that run. For example, the cheetah covers up to 7m per stride, which is nearly six times its chest-rump length. Cursorial adaptations are of real interest to the robot designer because it is difficult to make more than a few cycles of controlled movement a second with mechanical actuators; indeed, in many cases, it may take several seconds to move a leg through its complete cycle.

The most obvious way of increasing stride length is to use long legs. This only works if the stride rate, and therefore the angular speed of the leg, can be maintained. One of the factors affecting the average angular speed of the leg swing is leg inertia, so if the leg is to be made longer without slowing it down it must be made thinner. Therefore, cursorial animals have slender legs. The penalty is reduced strength.

The effective leg length can be increased in several ways. First, the foot can be extended so that it increases the length of the leg, with only the toes touching the ground. Such feet are called digitigrade. This principle is extended in animals that stand on the tips of their extended digits. This posture, found in hoofed animals, is called unguligrade. Second, the shoulder blade can be turned into an actively swinging and relatively long leg segment.

For a leg of given length, it is still possible to increase the stride in various ways. The first is to add the angular velocities of two or more joints. Instead of the leg swinging as a rigid member from the hip or shoulder, its segments all swing simultaneously about the joints at knee, ankle etc, with the result that the tip of the leg swings through a much greater angle than does any single joint. This is illustrated in Figure 3.2. Note that for multi-section legs, an additional cursorial adaptation allows the more distal segments to be longer and more slender than the more proximal ones. It is more important to reduce their inertia than that of the proximal segments because they move faster. For the same reason, the muscle masses are concentrated near the body and long tendons are used to actuate the more distal joints. There are other adaptations for reducing leg mass, but most are not relevant to robot design.

The angular range of each individual joint can be maximized (and therefore its angular velocity also maximized) by making the muscle

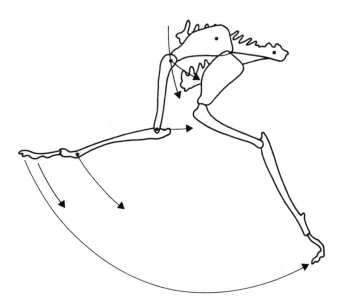

Figure 3.2 *Addition of angular velocities of several joints to increase the range of movement of the foot.*

insertion as close to the joint as possible. This is shown for a mechanical equivalent in Figure 3.3. Provided that the actuator can drive the increased load at the same absolute speed, it will produce a greater stride in a given period if it is attached nearer the joint.

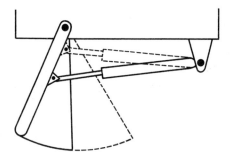

Figure 3.3 *The effect on the angular range, or stride, of the position of actuator insertion.*

Another way of increasing stride length is to use a flexible spine, or a jointed one. Lizards flex their spines in the horizontal plane so as to pivot the pectoral and pelvic girdles about a vertical axis so that the girdle becomes in effect an extra limb segment. Some mammals flex their spines in a vertical plane to alternately shorten and lengthen the distance between the front and hind legs, which in the gallop adds an increment to the stride. To quote Hildebrand (1974), 'The cheetah is so adept at this manoeuvre that it could theoretically run nearly 10km/h without any legs at all!' Spine flexing is in such a direction that the rotation of the girdles it introduces also adds to the reach of each leg.

Finally, in the running or hopping gait the distance the body moves forward while all feet are off the ground is added to the length of the step.

Adaptations for Economy of Effort

Some of these, such as reductions of the mass of distal parts of the leg, have been mentioned in Chapter 2, but animals have many subtle adaptations for this purpose, again described in Hildebrand (1974).

Miscellaneous Structural Features

Books on animal structure and function list many interesting mechanisms such as toggles, parallelogram linkages, ways of building cantilevers from bone and tendon, and so on. Some of these may be of direct use, or may inspire the designer to invent an analogous mechanism.

DIFFERENCES BETWEEN ANIMALS AND MACHINES

The preceding section suggests that the robot designer may learn much from the structure and mechanics of animals. However, the information so obtained must be treated with caution; a feature useful to an animal may not confer the same advantage on a machine. Therefore, this section lists some of the ways in which animals and machines differ.

1) An obvious difference lies in the type of materials used. Engineering materials are usually 'stronger' than biological ones, but rarely is tensile or compressive strength alone the most important property of a material; properties such as the work of fracture or the ratio of Young's modulus to density (E/ρ) may be more important under some kinds of load. To give two examples: although the tensile strength of mild steel is much higher than that of tendon, its tensile strength per unit weight is only two-thirds that of tendon; and the E/ρ of bone is about three-fifths that of aluminium alloy. This suggests that a structure made from bone and tendon may be as strong, in some ways, for its weight as a structure made from steel and aluminium.

Furthermore, biological processes can produce complex shapes and composite materials such as spongy bone, whose material distribution is very economical. On the other hand, it is possible to build a mechanical structure with much higher stresses at key points such as bearings because of the strength of materials. For more on this subject see Gordon (1978).

2) A biological structure can often limit stress by a combination of load sensing and muscular action.

3) One advantage of machines is that fatigue or energy consumption in the static support of a weight can be eliminated.

4) Animal components such as muscles often fulfil many roles at once, such as actuator, thermal insulator, cushion, support for other tissues and energy reserve.

5) A mechanical actuator may be heavier or lighter than a muscle. When the ancillary equipment such as valves, pumps and reservoirs is taken into account it is usually heavier. However, it may have some advantages: being able to push as well as pull; in some cases possess a greater extension ratio; the possibility of a greater force for a given cross-sectional area; and the choice of linear or angular movement.

6) The position and velocity control of rapidly accelerated coupled masses is much less highly developed for machines than in animals.

Leg Number and Arrangement

Walking (or hopping) machines have been built with one, two, three, four, six and eight legs and possibly more. Of these an even number of legs is almost universal as, for progression in a straight line, this allows efficient gaits. The higher numbers are suitable for heavily loaded slowly moving vehicles (for example in underwater construction), while bipeds and quadrupeds seem to be the fastest and most agile. Six is a magic number in that it allows two alternating tripods; and two is another popular number because of the interest in modelling and emulating human locomotion.

Some of the properties affected by the number of legs are:

1) stability,
2) energetic efficiency,
3) redundancy: the ability to use fewer legs if some are damaged,
4) quality of joint control required,
5) cost,
6) weight,
7) complexity of sensing needed,
8) possible gaits.

Not all these categories can be correlated with leg number. It seems likely that there will be tenable places for machines with several leg numbers.

By leg arrangement is meant the location on the body of the leg attachment points, the orientation of each leg and the internal geometry of each leg. A rigid body will be considered first but, as explained in the section on cursorial adaptations, a flexible or jointed body can improve locomotion. Bilateral symmetry will also be assumed.

Leg arrangement is more of a real issue for many-legged machines than for bipeds, which can only have their legs side by side or one in front of the other. The second arrangement would usually produce unacceptable interference between the legs and seems of little use.

The basic possibilities for four or more legs are shown in Figure 3.4.

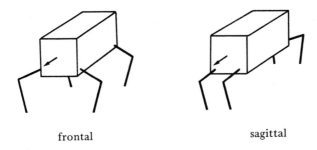

frontal   sagittal

Figure 3.4 *Frontal plane and sagittal plane leg arrangements.*

For the purpose of the discussion all legs are shown with just two segments and joined at the knee. For convenience, they will be referred to as the frontal and the sagittal pattern. The most fundamental distinction is whether the main plane of the leg is in the usual direction of motion (the sagittal case), or at right angles to it (the frontal case).

Within each category there are further design aspects to consider. First, is the geometry of each leg within its plane. As mentioned earlier, mammals tend to dispose the leg vertically below the body as is shown in Figure 3.5a; the reptilian arrangement is shown in Figure 3.5b; and an arrangement found in some arthropods is shown in Figure 3.5c. The very long but folding legs presumably confer high mobility on insects and spiders which are small compared with the typical dimensions of the substrate.

Another consideration is whether the legs at opposite ends (or sides) face inwards, outwards or the same way. (This is a crude simplification. Real animals do not have to fit neatly into these categories.) The configuration shown in Figure 3.6a has the least stability but there

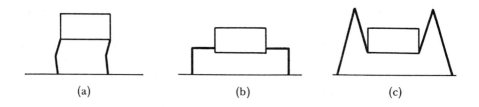

Figure 3.5 *(a) Mammalian; (b) reptilian; (c) insect leg dispositions.*

is also less change in the support pattern in changing from one set of legs to another; and it also allows standing on a small patch of firm ground.

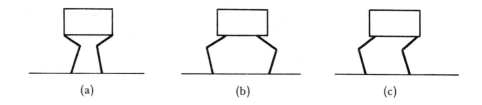

Figure 3.6 *The effect on stability and foot placement*
of the relative orientation of the legs.

For a machine with more than four legs it is important to know where to place the extra legs. They do not have to be in two rows, one down each side, like an insect. They could be placed in a row at each end, which is in effect the way a crab is arranged when it walks sideways. A machine with a row of legs at each end might have certain advantages, such as being narrower than the usual configuration. A heavy-duty machine might have many legs in an array covering its under surface, rather like the tube feet of sea urchins and starfish.

An additional consideration is the relevant proportions of the legs, which can affect walking in several different ways. For example, it may be desirable for the legs to overlap in order to eliminate interference, or to equalize their loading (see Figure 3.7). Such overlapping is often found in arthropods such as scorpions. In Figure 3.8 another example is shown of a hexapod using the alternating triangle gait. If the length of the middle leg is increased then the distance of the centre of gravity from the nearest edge of the base of support is also increased. A related rule states that the greater the ratio of vehicle length to stride length, the greater the stability.

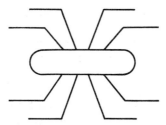

Figure 3.7 *Overlapping legs: the middle legs are longer than the end ones so they can make their stride without colliding with the end legs.*

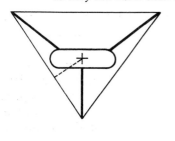 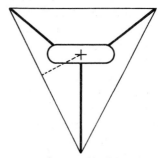

Figure 3.8 *The effect of leg length on lateral stability (only the three supporting legs are shown).*

Leg Design

There is no sharp boundary between leg arrangement and leg design. This section develops the previous one by going into greater detail on the possible geometries of legs and discussing some aspects of how legs can be designed to have various geometric characteristics. It is still at a fairly abstract level and is not concerned with engineering details. It lists some of the questions which must be answered, and in some cases expands on the approaches which have been taken:

1) How many degrees of freedom (DOF) are required, and how are these to be disposed?
2) Is foot orientation with respect to the body to be controlled (foot design)?
3) What type of linkages are required?
4) Should the method of varying the leg length be a hinge or telescopic joint?
5) Are cursorial or other adaptations required?
6) Are suspension elements such as tyres or springs required?
7) What type of servo control for the joints is required?
8) Are the joints backdriven, or elastic?
9) Is jumping, and therefore great power and speed and good energy recovery, required?
10) Is the leg to be equipped with other functions such as lifting, grasping or digging?

DEGREES OF FREEDOM: FOOT ORIENTATION

We can get a crude estimate of the number of DOF which ought to be needed by observing that during locomotion it must ideally be possible to control the six DOF of the body (three translational and three rotational) when it is supported by each of two alternating leg sets. So we might expect about 12 DOF to be a minimum for a walking machine. Of course, the leg sets may share some legs. There may also be fewer or more than two of them. However, if 12 DOF is taken as a rough estimate, they can be distributed among the legs in many ways, for example two legs of six DOF each, four legs of three DOF, or six legs of two DOF. In practice, we find both fewer and more than 12, but the number is generally of this order.

It is worth comparing the ways in which a biped standing on one leg and a hexapod standing on three legs control the position and orientation of the body. The single leg of the biped uses ankle roll and pitch together with leg length (which if the leg is not telescopic is determined by the knee angle) to control the position of the hip. Body orientation is controlled by hip pitch and roll together with yaw at either hip or ankle (see Figure 3.9a).

In the case of the three supporting legs of the hexapod, let us consider a simplified model of a leg which has two DOF, being length and angle from the body. Figure 3.9b shows that body position is a function of the differential length of the three legs, whereas body orientation is controlled by the three leg angle actuators. A real hexapod would need another DOF per leg so that it could move the raised legs about in a controlled way.

It can be seen that whereas the hexapod can have point feet, a biped must be able to control foot orientation, at least in roll and pitch. Yaw control can be provided at either hip or ankle. Of course, certain joints can be eliminated at the cost of not being able to control certain DOF of the body.

The leg with three DOF, which as we have just seen is necessary for hexapods, is also common for other leg numbers, so it is worth looking at how the DOF can be arranged. The most uniform mathematically are a Cartesian system of three sliding joints at right angles and a polar system in which a telescopic leg pivots about a two-axis rotary joint. The second design is used in the Carnegie-Mellon hopper. A more common arrangement is to use a knee joint. In this case, it is necessary to decide on the orientation of the leg and on which joints rotate about which axes. Figure 3.10 shows some possibilities.

The configuration of Figure 3.10a has often been used for hexapods, with the propulsion stroke being produced by rotation at the hip about the vertical axis. Those of Figure 3.10c and d resemble the geometry of crabs, and those of Figure 3.10e and f the geometry of some

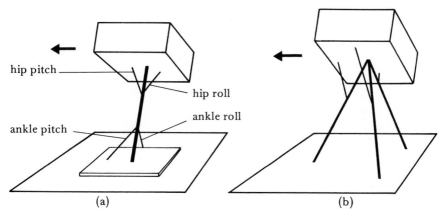

Figure 3.9 *Degrees of freedom of legs with and without control of foot orientation for (a) a biped; and (b) a hexapod.*

mammals. The last of these was used by the General Electric Walking Truck and also by the Ohio State University's Adaptive Suspension Vehicle (see Chapter 6). This subject is discussed further in the section on linkages.

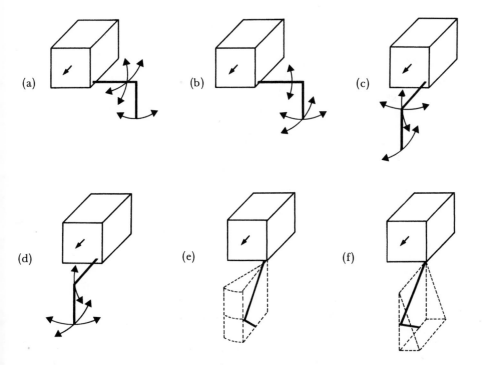

Figure 3.10 *Possible placements of joints for a three DOF leg.*

FOOT DESIGN

As explained earlier, a vehicle standing on three or more legs can control its body position and orientation even if the feet are regarded as points. Therefore, nearly all the work which has taken foot design into account has been that on bipeds. Nevertheless, when legged machines of all kinds are tried on soft ground the importance of foot design will become apparent.

A foot (the term as used here includes the ankle joint) can have several functions:

1) It must provide adequate traction, whether by friction or digging in or gripping.
2) It may provide a degree of resilient suspension by having a compressible tyre or sole.
3) If large enough to exert torque against the ground, it may be used to balance and orient the body. This may be done passively if the feet are very large, or actively if controlled ankle joints are fitted.
4) The foot profile may be designed to govern body height during the stride. For example, if the sole is shaped as a circular arc centred in the knee or hip, the foot can roll like a segment of a wheel, keeping body height constant.
5) It may have non-locomotory functions such as gripping or digging.

The feet of machines built so far have rarely if ever had more than two DOF at the ankle: roll and pitch. Changing the direction of walking is done using a yaw joint at the hip, or differential motion of the legs. A quadruped using the diagonal gait or trot could manage quite well with such a foot. Lateral stability would require good control of the ankle roll joints. So far they have tended to use very wide feet. (An example is the Phony Pony.)

The feet of wading birds with long toes show some potential for robots. They can be spread out to provide a wide base of support and folded up when raised to avoid interfering with the other leg.

LINKAGES

It is often possible to regard the main function of leg geometry as providing two main dimensions of motion: fore and aft, and up and down, with any displacement or, more usually, rotation of the plane itself being a secondary problem. The examples shown in Figure 3.10 can be regarded in this way.

In two dimensions, then, there is an endless variety of linkages by means of which two actuators can produce two more or less orthogonal

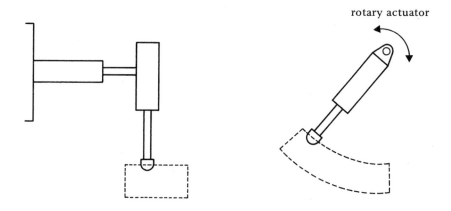

Figure 3.11 *Cartesian and polar leg arrangements.*

sets of trajectories of a point. As in the three-dimensional case the simplest mathematically are the Cartesian and polar arrangements (see Figure 3.11). These, however, are difficult to design satisfactorily, owing to the sideways force on the linear actuators. There are several criteria for choosing a linkage:

1) it may be desirable, in order to simplify control, to separate the functions of lift and propulsion,
2) some arrangements are dynamically or energetically favourable, minimizing accelerations or the mass of the reciprocating parts,
3) engineering: in particular, fluid cylinders are bad at supporting side loads, whereas rotary actuators are less readily available and imply large shaft torques,
4) the linkage should not interfere with the other joints needed for the third dimension of movement,
5) linkages can be used to add the displacements of pneumatic cylinders when these are used as two-position devices,
6) a parallelogram linkage can be used to maintain a link in a leg in a fixed orientation.

This discussion is limited to linkages having only two joints in the principal plane, but extra joints may be added, for example to allow a leg to fold into a smaller space.

The simplest linkage in common use is shown in Figure 3.12. Its advantages are its use of only hinged joints and its ability to use either rotary or linear actuators. There are many possibilities for its normal position (see Figure 3.10). Its disadvantages are the non-Cartesian foot movement and the non-linearity of forces as functions of joint angles. It can be modified to improve these characteristics. The example in Figure 3.13a has the advantage of keeping the orientation of the lower

rotary actuators

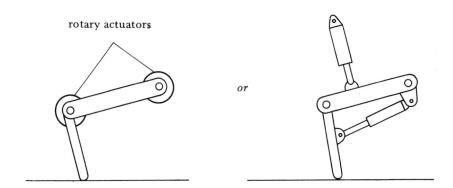

or

Figure 3.12 *Simple two-joint legs.*

section of the leg relatively independent of that of the upper section. The version shown in Figure 3.13b achieves this perfectly, at the cost of greater complexity.

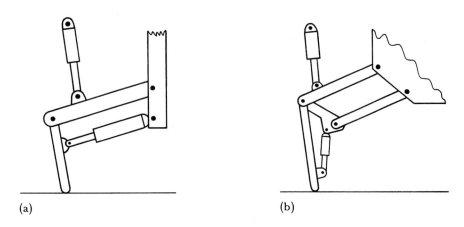

(a)   (b)

Figure 3.13 *Linkages for keeping the orientation of the lower leg segment independent of that of the upper section.*
This is achieved roughly in (a) but exactly in (b).

Perhaps the most common linkage for hexapods requires a modification of the planar model of leg movement. Rather than a plane, the tip of the leg is confined to a spherical surface. This is produced by rotation about a two-axis hip joint (see Figure 3.14). The knee joint affects the third dimension by changing the radius of the sphere, and could be replaced by a telescopic joint. This linkage is also shown in Figure 3.10a.

Turning now to more complex linkages, some have been devised to

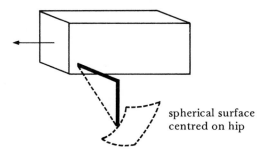

Figure 3.14 *Three DOF leg using two-axis hip joint.*

produce straight-line motion without imposing side loads on cylinders. An early example of this kind of linkage was invented by James Watt and is found in beam engines. One of the leg designs by Hirose and Umetani (1980) has the property of generating straight horizontal motion using rotary joints connected by wires and pulleys. This is shown in Figure 3.15a.

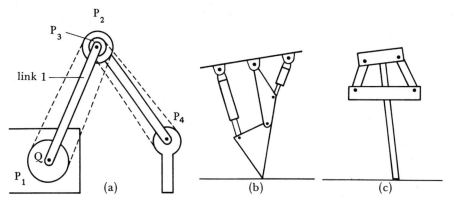

When P_1 is fixed to the body and link 1 rotated about Q
the foot moves horizontally. If P_1 is locked to link 1 instead,
the foot moves in an arc centred on Q.

Figure 3.15 *Linkages producing straight horizontal motion of the foot.*

The linkage of Waldron and Kinzel (1983) (see Figure 3.15b) also produces nearly straight horizontal motion over a useful range when actuator B is locked. Another linkage which produces nearly straight horizontal motion is shown in Figure 3.15c. By varying the proportions of its links the foot trajectory can be made to curve upwards or downwards.

Pantographs have been used to transfer straight-line motion from one part of a machine to another, and to amplify the motion, but do not in themselves eliminate side loads on the actuators (Shigley 1961; Hirose and Umetani 1980). The use of a pantograph for this purpose, and the method of dealing with side loads, in the Ohio State University's Adaptive Suspension Vehicle is described in Chapter 6, which also includes other examples of pantographs.

The linkages shown in Figure 3.15a and b have energetic advantages in addition to lift/propulsion decoupling. In Figure 3.15a if the locked state of the pulley is suddenly changed from 'body' to 'link 1' the kinetic energy of the horizontal body motion will be conserved over the transition and converted to kinetic energy of motion along the curved (possibly nearly vertical) trajectory. In Figure 3.15b actuator B acts across two joints, as often happens in animals. It is attached to the body, so it does not participate much in the energy-wasting oscillation of the leg.

RECTILINEAR JOINTS

A rectilinear joint is one in which the moving parts describe a straight line. There are several types.

A telescopic joint is one in which one cylindrical component slides inside another and, in practice, it invariably derives from the use of a hydraulic or pneumatic cylinder. A sliding joint can have a flat geometry like the saddle/bed joint of a lathe or it may be telescopic. Alternatively, a rectilinear motion may be produced by a nut riding on a lead screw, a mechanism used in, for example, the Odex I. A telescopic joint may contain several concentric cylinders to increase its extension ratio. This process of compounding may be applied to non-circular cross-sections, as in the jib of a mobile crane. The term *telescopic* as used here assumes a circular cross-section unless qualified. A telescopic joint can rotate freely about its long axis, so if it is required to resist torsion it is necessary to mount two or more side by side. This can be seen in the largest Unimate industrial robots. If a non-circular cross-section is used to resist torsion, then conventional fluid actuators cannot be used. They are always cylindrical, except for rotary ones, because manufacture and sealing are much simplified.

Some sliding joints can resist heavy transverse loads, but many fluid cylinders cannot, at least when fully extended. Therefore, sliding joints are rarely used on legs. However, they can be successful in suitable applications. The design of the sliding joints of the Ohio State University's Adaptive Suspension Vehicle, which uses conical rollers and V section guideways (described in Chapter 6), shows how elaborate the engineering of sliding joints to carry heavy loads may become.

Structure

Comparatively little attention has been paid to the structural aspects of walking machine design. Yet the designer must face questions such as whether the main structure should consist of a narrow spine, a space frame or a monocoque, what materials to use and so on. As the new generation of high performance legged robots is developed, these issues become more important, and more effort is now being put into structural design (Vohnout *et al.* 1983; Hirose *et al.* 1984). This section draws attention to some particular problems of walking machine structures.

TORSION

A machine with a long horizontal axis such as a quadruped will sometimes be supported by legs that are widely separated and on opposite sides of the machine. This produces a large torque about this axis (the spine, if it has one). Designing a structure to resist this torque is much harder than merely making it strong enough to resist bending, tension and compression. The problem is worse for a reptilian geometry than for a mammalian geometry, as the legs are further from the centre line and the torque is, therefore, greater. For a vehicle of a given length the problem is less severe for an insect-like hexapod because its legs are half as far apart along the torsion axis (see Figure 3.16).

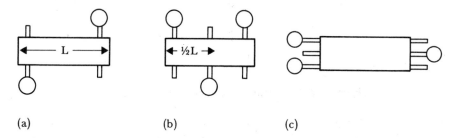

(a)   (b)   (c)

supporting legs are marked with an open circle

Figure 3.16 *The effect of leg arrangement on body torsion.*

Figure 3.16c shows that a hexapod with three legs at each end instead of three on each side introduces no torsion about the long axis, at least in the alternating tripod gait.

Torsion also presents a problem in the leg, particularly in the upper section or femur. The section below the knee can act as a lever to turn propulsive thrust or the vehicle's weight into a torque about the long axis of the femur. The magnitude of this torque depends on the geometry, on the slope of the ground and on the acceleration of the vehicle.

MISCELLANEOUS STRUCTURAL CONSIDERATIONS

This section lists an assortment of factors which need to be taken into account. The first of these is the influence on design of the need to dispose various bulky objects about the vehicle. For a multi-legged robot it is usually desirable to keep the centre of gravity as low as possible so that the machine will not overturn on slopes. This is not much of a problem with a low wide machine but is important for one intended to pass through narrow openings. Therefore, heavy components such as engines and hydraulic valves should be placed as low as possible in the system. One way of doing this is to sling them from a high spine. This is particularly appropriate if the legs swing in an almost vertical plane, as they can have short attachments to the spine. The low-and-wide and the tall-and-narrow approaches are shown in Figures 3.17a and b. Of course, these arrangements may be modified. For example in Figure 3.17c the spine is replaced by a box, which is better at resisting torsion. Its sides may be filled in sheets or diagonally braced frames.

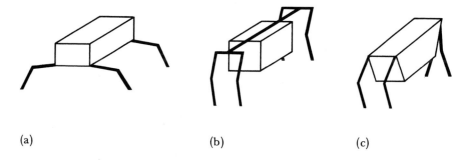

(a)   (b)   (c)

Figure 3.17 *Alternative structures for a quadruped.*

Another consideration is the provision of large open volumes for bulky components such as engines, which also allow access. It is all too easy to find that the inner parts of the machine are surrounded by an impenetrable thicket of legs.

If the vehicle is to be manned, then it will influence the design, mainly because of the need to provide a safe comfortable compartment with adequate visibility. If the driver can control individual legs, then visibility is more of a problem than it is for wheeled vehicles where, except in very large systems, the driver can usually ignore the exact path taken by the rear wheels. Two solutions can be seen in the General Electric Walking Truck, which was fitted with big rear-view mirrors, and the Carnegie-Mellon hexapod, where the driver sits at the back and has a clear view of all the legs.

STRUCTURAL PROBLEMS OF BIPEDS

Bipeds may be divided into human exoskeletons and autonomous robots. These classes have some problems in common and some of their own. A common problem results from the need to balance on one foot; the biped leg must resist bending and torsional stresses which are much smaller or absent in the multi-legged machine with point feet. A related problem is the weight and complexity of the ankle mechanism. The result is that the legs of bipeds, even when not exoskeletons, usually require a space frame or a large diameter tubular construction.

Exoskeletons obviously have to conform to the shape of the human body. It must be possible for the wearer to get in and out of the suit. In the case of Vukobratovic's exoskeleton each leg segment was made of semicircular metal bands round the back of the wearer's leg, connected to rods running parallel with the leg, with straps fastening the circular bands to the leg. In the General Electric Hardiman (see Figure 1.5) the main structure runs parallel with, and on the outside of, each of the wearer's legs. This is massive compared with the orthotic exoskeletons, which have to be less bulky.

STRUCTURAL PROBLEMS OF LEAPERS

Figure 3.18 shows how a one- or two-legged hopping or leaping machine experiences large bending forces, particularly on landing, and suggests the sort of braced structure which is suitable.

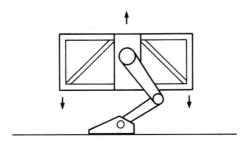

Figure 3.18 *Stresses on a leaping machine.*

The Carnegie-Mellon hopper has diagonal bracing of this kind, although it is circular in plan. A machine which can run fast on two legs would also experience high bending forces in the leg and its attachment, and shear across the body as shown in Figure 3.19. The stresses on the leg bearings are also very great.

Actuation Methods

This section is a superficial introduction to the engineering of hydraulic,

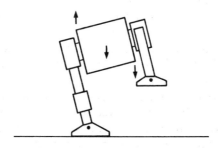

Figure 3.19 *Bending and shear stress on a biped.*

pneumatic and electric actuation. It also suggests some comparisons between them, although no universally valid conclusions can be drawn, partly because special circumstances attend every design, and partly because of the rapid rate of development in these areas.

First of all, why limit the discussion to these three kinds of actuation? Others are certainly possible, for example all-mechanical transmission (of which a brief account is given later in this section) or the use of piezoelectric actuators. However, at present the first three means offer overwhelming advantages in most applications. This is simply because they all have a long history of use in power transmission, with which is associated a massive amount of knowledge and available components. The robot designer who uses one of these methods can take the underlying technology as given, although he may well find it inadequate in some respect and become involved in the search for a better component.

Mechanical transmission is also a developed technology. Its failing is not that it is an untried technology but that, for the present at least, it does not seem well suited to the particular characteristic of legged locomotion, which is the need to power many joints, some of which are at the end of jointed structures which would make the use of shafts, gears, belts and so on very cumbersome. The situation may change as new devices are invented and new materials developed. After all, animals can transmit complex motion across several joints. (Consider the tendons of the human fingers, for example.) However, ever since the Iron Mule Train and other similar machines of the time, most robot designers have abandoned mechanical transmission.

ELECTRIC ACTUATION

There are many types of electric motor which could in principle be used as joint actuators. A way of classifying them is to distinguish between motors which produce a large torque at low speeds and can be connected to the joint directly, or nearly so, and those which run at high speeds and need a large gear ratio to produce a low speed, large torque output. Examples of motors in the first class are d.c. and a.c. servo motors. They are common in industrial robots, and servo motors have been used in small bipeds (Miura and Shimoyana 1984). The second class consists of all those motors which run at speeds of up to a few thousand rpm and whose speed and direction can easily be controlled. One example is the ordinary d.c. motor; another is the electric drill motor used in the Ohio State University's hexapod. This class needs a light and efficient gear train. Worm gear boxes and lead screws have been used; a high performance alternative is the harmonic drive. With high gear ratios (of the order of 100 to 1), and especially with worms, the joint is often not backdriven, which may or may not be desirable in a particular application.

The main limitation of electric drives is that their power-to-weight ratio is poor, particularly if an on-board generator is included. This limitation does not matter for industrial manipulation robots. Future high performance batteries or fuel cells might change this, but at present electric transmission seems limited to fairly slow walking.

HYDRAULICS

One of the chief attractions of hydraulics is its ability to generate large forces directly, avoiding the need for gearing. This is particularly pronounced for very large machines, and the largest presses, mobile cranes and so on are usually hydraulic. It may be noted in passing that in these applications the actuator need have only a finite stroke; where infinite travel is required as in driving the wheels of a vehicle hydraulic transmission, although sometimes used, it is at less of an advantage. A second attraction is its high force-to-weight ratio (and power-to-weight ratio), which results from high working pressures, commonly of the order of 140 bar (2000 psi). A hydraulic motor can in some cases weigh a tenth as much as an equivalent electric motor. Other attractions are the stiffness of hydraulic drives and the lack of a need for a separate braking mechanism which results from the fact that, by simply closing a valve, a cylinder can be turned from a backdriven condition to a rigidly locked one.

A standard hydraulic circuit is shown in Figure 3.20. The pump may give a constant flow rate at a given speed, a common example being the gear pump; or it may give a controllable flow. Such a pump, called a variable displacement pump, has a means of varying its swept volume in response to a hydraulic, mechanical or electrical signal. An example is the swash plate pump in which the stroke of a set of pistons in cylinders mounted round and parallel with a central shaft

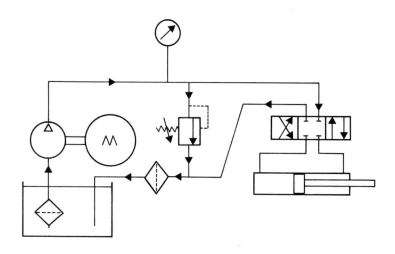

Figure 3.20 *A standard hydraulic circuit.*

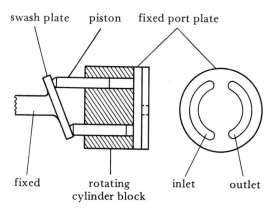

Figure 3.21 *A variable displacement pump;*
the angle of the swash plate controls the stroke and hence displacement
of the axial pistons.

(like the chambers of a revolver) is varied by the angle of a plate against which their ends bear (see Figure 3.21).

In the circuit shown in Figure 3.20 the load is a cylinder, or ram as they are often known, controlled by a spool valve with three positions. In the centre position the valve isolates both ends of the cylinder. When the spool is driven to one end by a solenoid or otherwise, one side of the piston is connected to the pump and the other to the return line. With the spool at the other end, the reverse connection is made. In the example the valve is either fully open or fully shut, but versions exist where the flow can be accurately controlled at any intermediate value. These are servo valves and are very expensive. An intermediate type, offering a flow in some degree proportional to the solenoid current, is also available. The load could be a rotary actuator, of limited angular range, or a motor, instead of a ram.

If a variable displacement pump is used the speed of the load can be controlled without using valves. Such a system is called hydrostatic; the force input to the pump is transmitted via fluid pressure to the force output of the actuator. The pressure is not determined by valves. It has the advantage that no power is dissipated in the valves. An example of a hydrostatic circuit is shown in Figure 6.4. If it is required to control the speed of several actuators separately, then there are three possibilities:

1) A common fixed delivery pump, with a servo or proportional valve per actuator.
2) A variable displacement pump for each actuator and no valves.
3) A small number of variable displacement pumps, with a directional but not proportional valve for each actuator. This does not allow the speed of all actuators to be controlled completely independently but divides them into groups, the speed of each group being independent of the others.

The circuit shows the other essential elements of a hydraulic circuit: the oil reservoir (oil, quite similar to engine lubricating oil, is the usual working fluid, although water-based hydraulic fluids are also known), inlet strainer, pressure gauge, safety relief valve and return line filter. In addition, an oil cooler and an accumulator consisting of a gas volume separated from the oil volume by a piston or flexible diaphragm may be fitted as a way of damping fluctuations in pressure resulting from changes in the load. For details of these elements and their use see works such as McCloy and Martin (1980).

As applied to legged robots, hydraulic actuation has a few features to be noted. First, it is not suited to applications requiring spring-like behaviour, as in hopping, unless the actuator can be closely coupled to an accumulator. Second, it allows a simple method of obtaining a degree of conformation to stairs or rough ground; the legs of a set are lowered and as each touches the ground it is stopped until all are touching, when the legs can all be locked rigid or all extended an equal amount further.

PNEUMATICS

The standard pneumatic circuit, shown in Figure 3.22, is very similar to the hydraulic circuit. A return line is not needed as the actuators vent to the atmosphere. The cylinder control valves are usually on/off ones; proportional or servo control valves are unusual. A non-return valve is fitted between the compressor and the reservoir so that compressed air is not lost when the compressor stops. Small compressors use pistons or rotating vanes; larger ones sometimes use a screw.

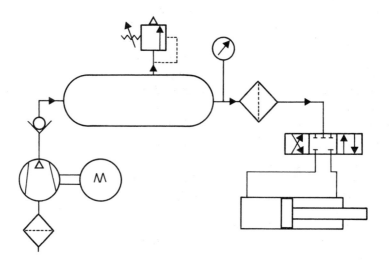

Figure 3.22 *A standard pneumatic circuit.*

Gas turbines use axial flow or centrifugal compressors, which must run at very high speeds to be effective. Pneumatic systems are economic and easy to assemble; simple nylon tube and push-in fittings can be used at the usual pressure of 7 or 14 bar (100 or 200 psi), there is no oil to leak and no problem of air bubbles getting trapped in the system.

A fundamental difference from hydraulic systems is that, since air is an elastic medium, a pneumatic actuator cannot usually be treated as rigid. This implies that servo control of position is difficult or impossible, particularly if the load fluctuates, as the actuator will behave as a spring. Therefore, most pneumatic machines use actuators which drive between their end stops at a more or less constant rate determined by the load and the air flow rate, and do not stop between the ends. Hence, they may be regarded as two-state devices. This being so, it is possible to compound the stroke of several actuators using a linkage. Such compounding can also be used to select one of several speeds. An example of the use of this method is the pneumatic quadruped of Taguchi *et al.* (1977).

Pneumatic servo systems

Although pneumatic cylinders are usually operated as two-state devices, position and velocity servo control are possible under some circumstances. For it to be practicable the pneumatic force available must be high compared with friction and inertia, and there must be an adequate air or gas supply. The first condition usually requires high pressures. (If the actuator is thought of as a spring, its stiffness is proportional to the pressure; therefore, in principle, a pneumatic servo can be made indefinitely stiff by increasing the pressure.) An example of high pressure pneumatics is actuation of the control surfaces on missiles, using filtered combustion gases at pressures of the order of 100 bar. The components are not readily available at present but this approach cannot be ruled out for the future.

Pneumatic servo systems at the more usual pressure of 7 to 10 bar (100 to 150psi) are used in some industrial robots for fairly light loads. The positional accuracy is improved by the use of pneumatically operated disc brakes. This approach seems to be of little use for walking machines where the loads are much greater.

The leg of the Carnegie-Mellon hopper is pneumatic and is part of a servo system but the controlled quantity is not actuator position or velocity, and control is not continuous. For an account of its operation see Chapter 4.

OTHER FORMS OF ACTUATION

Internal combustion

This concept was also suggested independently by G.H.D. Darwall. Observation of hand-controlled machines for levelling road surfaces by impact (for example the 'Wacker' often seen during road repairs in Britain, which leaps to a height of about a foot between impacts) suggests the possibility of a leaping or running machine whose main propulsion thrust is generated from internal combustion in the leg actuators. This would have the advantage of generating mechanical energy where it is required, with a greatly reduced need for engines, generators, oil pumps and so on. Although at first sight a rather violent and uncontrollable technique, the history of gasoline and diesel engines suggests that internal combustion actuators for legged locomotion could eventually be made smooth and powerful. As illustrated by horses and kangaroos, fast running or jumping is not suitable for heavy freight or comfortable passenger transport, but may be the fastest form of travel for a rider, or for an autonomous machine, crossing rough country.

Some issues which would need to be investigated are listed below:

1) How is thrust to be regulated throughout the stroke? The most obvious way is by controlling the rate of fuel injection.
2) Can an effective intake-combustion-exhaust cycle be devised?
3) Is it possible to operate by internal combustion at the low speeds needed for walking slowly? Or is a hybrid design possible, in which at low speeds the actuators work as conventional pneumatic devices powered by a relatively small on-board compressor?

An internal combustion locomotion machine would probably still use another kind of actuation, for example hydraulic, for everything except main propulsion.

Mechanical Transmission

Gears, levers and so on can be used with any form of actuation to transmit force or torque from an actuator to a joint. However, here mechanical transmission is defined as the use of clutches and other mechanical components to allow *more than one joint* to be driven from a single motor (typically an internal combustion engine). As explained earlier, at present it is less satisfactory than electric or fluid power, but a brief discussion is in order.

First, it may be remarked that although some early walking machines such as the Iron Mule Train were all-mechanical they did not allow independent control of the joints powered by a single motor and, therefore, could not adapt to the ground profile.

A walking machine which does come within the definition (just) used here is the Kumo-1 (Hirose and Umetani 1980) in which each of the four leg motors drives two joints, the joint to be powered being selected by locking a brake. As explained earlier, this machine has interesting energetic properties. It also shows how mechanical transmission can sometimes reduce the mass of the legs by concentrating the heavy actuators near the body. In its use of four motors rather than one it is clearly not an example of purely mechanical transmission. Also, each motor is not able to drive a third joint, which would have made a more convincing example.

At least one manipulator has been built with seven joints driven by a single motor, using clutches and brakes. This shows that the problem of geometric complexity can be overcome. However, only one joint could be powered at a time. This is not inevitable, but a mechanical transmission as versatile as an actuator-per-joint system would need a variable speed drive for each joint, which at present seems rather too complex to be attractive.

Chapter 4

The analysis and control of legged systems

Purpose of Analysis and Control

The analysis of legged locomotion is interesting in its own right in zoology and medicine. From the point of view of the robot engineer, analysis is seen mainly as an aid to design.

There is at present no systematic way of designing a legged robot to meet a performance specification. The designer usually starts with a particular idea he wishes to investigate, or a particular leg arrangement which seems to make stability easy to achieve, and from this the rest of the design follows. Once a basic approach has been chosen, it becomes necessary to analyse the kinematics or dynamics of the machine and design a control system. Of course, the reason for choosing a particular design may be to investigate analysis and control issues, in which case the designer will ask himself what mechanical arrangement will be best for testing a certain aspect of analysis or control.

The result of this haphazard process is that, at present, there are only analyses for those designs which have been thought interesting and none for legged machines in general (except in limited respects such as the number of gaits possible for various numbers of legs).

This chapter describes several techniques which have been found useful for various types of vehicle, but it begins with an account of the objectives of a robot control system.

The Objectives of Control in Legged Locomotion

A legged vehicle is a servo mechanism with many degrees of freedom. It can usually be approximated by a series of rigid masses connected by hinges. Each joint may be sprung, damped or powered by an actuator. Control in this context means finding a set of actuator forces, varying over time, which will result in *sustained stable motion* of the vehicle in a specified direction. Sustained stable motion can be broken down into several objectives, of which the following are the most important:

1) stability, that is, not falling over,
2) maintenance of body orientation,

3) control of forward velocity,
4) avoidance of vertical jolts,
5) ability to turn and reverse,
6) ability to do all the preceding on rough ground (this is what is meant by *terrain adaptability*),
7) ability to cross or pass between obstacles.

Of these, 2, 3 and 4 are related aspects of the general control problem and will not be discussed separately. Stability, terrain adaptability and dealing with obstacles, however, are sufficiently interesting to warrant more detailed attention.

STABILITY

One of the main considerations in legged locomotion is how to achieve stability, that is, avoid falling over. This is a potential problem for all legged animals or machines, as the constantly changing shape and variations in the slope of the ground are likely to result in unstable states. An unstable state is one in which the resultant vector of the gravitational and inertial forces meets the ground outside the base of support. Some examples are shown in Figure 4.1.

Stability must be treated somewhat differently in the case of standing still, where only one stable state is involved, and in locomotion, when stability is more a matter of achieving a stable cycle and may or may not involve passing through states which are stable in themselves.

Stability of standing can be achieved in two ways. The simplest is *static stability* and occurs when, as shown in Figure 4.1a, there are no significant accelerations and the arrangement of legs in contact with the ground is such that the centre of gravity is always above the base of support, which is defined as the convex hull of the polygon connecting its feet. This can be done by, for example, always keeping at least three feet on the ground. No movement is needed to maintain balance. Alternatively, if the machine has a very small base of support, servo control of the joints can be used to damp out any disturbance and balance the system. This is only possible with certain configurations.

During locomotion the simplest form of stability is to pass without a break from one stable state to another. In this case the machine can move arbitrarily slowly and stop at any point. (The stability of each state it passes through can be achieved statically or by balancing.) Other machines, however, pass through stages in the locomotion cycle which are not inherently stable. In effect, the machine is temporarily falling. In this case a different kind of stability must be sought. It is called *gait stability*, and applies when a periodic motion occurs in which, although the centre of mass of the machine may oscillate, on average it remains at a constant height. It could be described as a limit

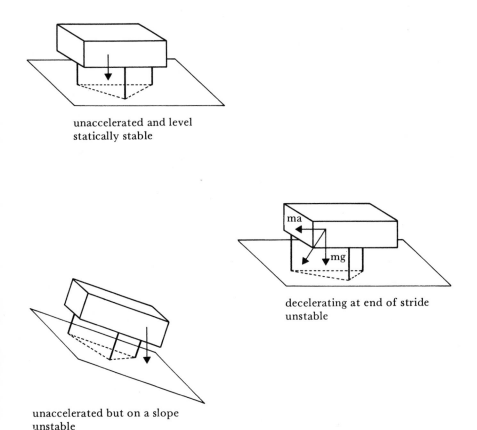

Figure 4.1 *Causes of instability in a multilegged robot.*

cycle in which feedback controls the amplitude and period of the oscillation.

If the gait does not contain phases which are stable in themselves, the machine will not be able to stop without falling over, unless it changes gait. An example is the Carnegie-Mellon hopper; as it has a point foot and cannot exert a moment against the ground, it is not stable when stationary and, therefore, must keep hopping.

The terminology is not used consistently throughout the literature; for example, passive stability has been used to denote the term referred to here as static stability, but also for those forms of gait stability which include some stable and some unstable states. The latter will be referred to as *alternating balance stability*. The different forms are summarized in Figure 4.2.

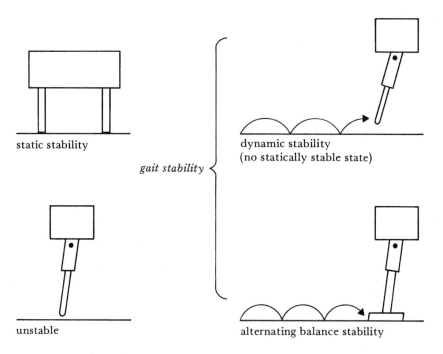

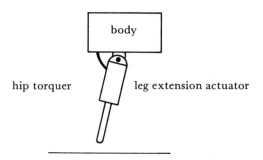

Figure 4.2 *Definition of different kinds of stability.*

An interesting feature of gait stability is that it makes use of degrees of freedom of the system which are not fully controlled in the sense of conventional servo control. For example, in the planar hopping machine shown in Figure 4.3 the system has five DOF (body position in two coordinates; body angle, leg angle and leg extension), but it only has two actuators so it cannot control all five variables completely. Nevertheless, stable locomotion is possible because the system is partially constrained, mainly by inertia. Once per stride all quantities

Figure 4.3 *A planar hopping machine which can achieve a stable cycle with just two actuators.*

can be restored to, if not a precise nominal value, at least one which allows a regular oscillation to continue.

Most animals change from statically stable gaits at low speeds to more dynamically stable ones at high speeds. In designing a vehicle it is much more difficult to achieve dynamic stability than static stability; the servo systems controlling its joints must be fast enough to keep the machine balanced. Therefore, many walking machines have been designed to be statically stable. The most common way of doing this is to provide six legs and move them in triplets so that the robot is supported by two tripods alternately. Static stability can be achieved with fewer legs provided the feet are large enough to define an adequate base of support. For example, bipeds with large feet can walk slowly in a condition of static stability by keeping the centre of gravity always over one foot or the other or over the area between them when they are both on the ground.

At high speeds static stability must be abandoned. All high speed gaits (running, kangaroo hopping) require the animal or robot to leave the ground periodically; and even in the brief intervals of ground contact only one or two legs touch the ground at once.

TERRAIN ADAPTABILITY

When a robot must walk on ground which is not smooth and level several control problems may arise:

1) navigation: choosing a route at the most global level,
2) path selection: that is, given a roughly specified route, choosing its details so as to minimize the problems due to slope, roughness and obstacles,
3) terrain adaptation and obstacle crossing.

Of course, these levels are not always distinct; for example, there may be a choice between putting a lot of effort into choosing a path which avoids obstacles and putting effort into dealing with obstacles. This section discusses terrain adaptation, which to a large extent can be treated separately from the problem of negotiating obstacles such as ditches or fallen trees.

The problems of terrain adaptation are rather different for a multi-legged robot and a dynamic one which runs or hops, landing on only one foot at a time. In the second case, ground slope does not necessarily affect body attitude. The control problem is largely one of selecting foot placements to minimize jolting and the danger of slipping. No attention seems to have been paid yet to this case.

For a robot with multi-legged support terrain adaptability means selecting the positions of foot placement and changing the effective leg

lengths to meet the following objectives:
1) minimizing fluctuations in body path and orientation in order to save power and give a smoother ride,
2) keeping body orientation independent of ground slope,
3) enabling the robot to walk on steeper slopes than would otherwise be possible (see Figure 4.4).

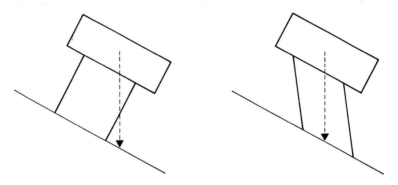

leg-body angle fixed: leg-body angle adapted to slope:
unstable   stable

Figure 4.4 *Example of terrain adaptability:*
adjusting mean leg angle to keep body centre of mass
within the base of the support.

Terrain adaptability is not always necessary; the machine may not be required to walk on rough ground, or it may not matter if it does not move smoothly.

Adaptability imposes certain sensory and actuation requirements. It must be possible to measure joint angles, foot-ground contact and body roll and pitch angles. Servo control of the joint actuators is required, and their range of operation must be great enough to allow an offset to the walking stroke.

OBSTACLE CROSSING

Passing between obstacles where possible is outside the scope of this section; it is an aspect of path selection. By obstacle is meant a step, wall, trench, post, block or hole low enough to step on to or narrow enough to step across. Obstacles larger than a certain height and width cannot be crossed. For a statically stable walker this limit is about the same as the length of the stride and the vertical travel of the foot (unless the machine is very long, like a centipede). For a runner or leaper it is determined by the length and height of the leap or hop.

Hirose (1984) has suggested a classification of terrains according to whether they contain post-like or hole-like obstacles (types P and H, respectively), both (type HP) or neither (type O). His last type, G, is a general terrain model, divided by a regular lattice into cells each containing a flat surface. This can be reclassified as O, P, H or HP depending on the relative height of neighbouring cells.

The rest of this chapter is a survey of what is known about the analysis and control of legged machines. It is divided into two sections for convenience of presentation, although these areas overlap.

The first section covers dynamics; in particular, the application of control theory to linkages. This is relevant to all legged robots, but the emphasis is on bipeds, since it is in bipeds that dynamic issues have been most crucial.

The second section deals with matters such as gait selection and foot placement, some of which are often treated as problems of kinematics rather than dynamics. Here the emphasis is on multipods. This is so partly because they face the problem of coordinating many legs and partly because, at present, research on multi-legged robots regularly addresses the problems of rough ground and obstacles, whereas the work on bipeds has largely ignored them. The problems of achieving stable locomotion are so severe that most research has been confined to their behaviour on level ground. Of course, there are exceptions to these generalizations.

Dynamics

The dynamics of a legged robot can be modelled in terms of a linkage of rigid units connected by hinged or sliding joints. There is an extensive literature on this subject. The purpose of this section is to outline the approaches used, and to explain some of the terminology. It begins by discussing the aims of dynamics as applied to legged locomotion.

AIMS OF CONTROL THEORY

Aim 1

The most important aim is to find the joint torques or forces which must be generated for the machine to produce a specified pattern of movement. A control law has to be found which gives torques as a function of the state and desired behaviour of the system. This is the *design* or *synthesis problem*.

The required output may be specified in several ways:

1) a set of prescribed joint angles as a function of time,
2) a prescribed motion of certain key parameters or variables
   such as centre of mass and body angle,

3) a path which optimizes some quantity such as energy or acceleration, while still meeting some specification such as the position at a particular time,
4) stability of the machine against disturbances.

Note that the design problem can be subdivided into four stages:

1) formulation of the equations of motion in terms of the unspecified joint actuator torques,
2) solving these equations to give a set of relationships between the joint torques and the coordinates (typically the joint angles),
3) specifying the required output,
4) finding a control law which when substituted into the equations resulting from 2 (above) will give the desired behaviour of the coordinates.

Aim 2

The converse problem is to find, theoretically, the motion of joint angles or other spatial variables, given a set of torques as a function of time as the input. The ability to solve this problem is important for various methods of analysis and control.

Aim 3

Another aim of control theory, which applies more to animal and human locomotion but can still be relevant to machines, is to observe the motion of a mechanical system and model it. The model usually takes the form of a set of rigid links or masses linked by joints about which there may act torquers, springs or dampers. It is often not possible to observe all the relevant variables. Some, such as joint angles, may be measurable in principle but difficult to instrument in practice. Quantities such as the position of the centre of mass cannot be measured directly and must be inferred. Further, there may be very many possible variables or coordinates, and it is necessary to choose a set which characterizes the system properly.

APPROACHES TO CONTROL

Some approaches to analysis and control will now be discussed. A useful distinction is between attempts to deal with an essentially *complete system* (system refers to the mobile, that is the animal or machine, together with the ground) and *reduced order models*. In these models the problem is simplified by reducing the number of degrees of freedom of the model by, for example, treating the motion in only one plane at a time, or reducing the number of links in the model.

Of course, a complete model can never be achieved perfectly, but in the case of a robot a 'complete' model can include all the major components and need make no assumptions about motion being confined to a subset of those physically possible. The reasons they are not the only ones used are their need for efficient computational methods for formulating and solving the equations of motion, and the fact that even quite simple reduced order models provide a rich field for investigation, which is by no means exhausted.

A second useful set of concepts in classifying control methods is that of *finite state control* and *model reference control*. (See, for example, Hemami 1976; McGhee 1977.) Suppose we wish to find a joint torque control law to produce a prescribed trajectory of the joint angles or other quantities (aim 1 as just defined). The method of solution depends on how accurately, and at how many points along the cycle, we wish to specify the output state. (The trajectory will be a closed path of finite duration, since locomotion is a cyclic activity.)

Finite state control refers to the case in which it is sufficient to stipulate that the system goes through a cycle of a finite number of states at the end of which it has moved forward. It has several interpretations. Tomovic (1961) modelled the locomotion of earthworm-like mobiles in which two-state actuators between the segments alternately expand and contract. He modelled the dynamics in terms of impulses. Finite state control has also been used to refer to the purely kinematic description of systems which execute a cycle of states, ignoring the dynamics. This approach is suitable for slowly moving vehicles where inertial forces are negligible. An example is the Phony Pony in which each of the four legs used a finite state controller. Each of the two joints in each leg had four states: free swinging, locked, powered forward, powered rearward (see Chapter 1). A simple behaviour which may be regarded as finite state control is the limit cycle in which a machine rocks back and forth between two states (which do not have to be stable in themselves). In other treatments the system passes through a cyclic series of equilibrium states in each of which the system must be stabilized dynamically. In this case we speak of finite state control 'supervising' some other form of control which is responsible for stabilizing the equilibrium states.

Model reference control is, in one sense, the opposite of finite state control in that the trajectory of all joint angles is completely specified at all times rather than being determined only at intervals, but as will be shown, hybrid forms of control are possible. In model reference control all joints are servo controlled (using the joint torque actuators) to follow a nominal angle-versus-time function, called the *reference trajectory*. This, for an n-joint system, can be thought of as the history of the n-dimensional vector of joint angles; it is a curve in n + 1 dimensional space, the n + 1th dimension being time. But it is often convenient to speak of the reference trajectory of a single joint, which is simply its specified angle as a function of time. The reference trajectory can be derived:

1) by observing, for example by filming, a person or animal,
2) by recording the joint angles of a model which is moved through the locomotion cycle of positions,
3) from a mathematical model (simulation).

In the last case only, it is possible to generate the reference trajectory in real time, if the simulation is fast enough. This allows adaptation to unforeseen changes in, for example, the ground profile.

In the purest form of model reference control the reference trajectory is a continuous function of time, for each joint. However, it may consist of a series of abrupt steps separated by constant intervals. This is to be interpreted as a series of static equilibrium points to be achieved in turn. The time at which each step occurs may be fixed by a clock or by some controller which uses some measure of the condition of the system, for example how closely the reference angles are being achieved. A system which behaves like this can be seen to conform to the description previously given of finite state control supervising some other form of control; the 'other form of control' in this case is model reference control. An example is given on page 113.

Given a reference trajectory, how should the joint torques be determined? There are two approaches. In the first, the angle (θ) (and its rate) of each joint is simply fed to a servo of, for example, the form:

$$M = k_p(\theta - \theta_{goal}) + k_v\dot{\theta} \qquad (4.1)$$

where M is torque, θ is the actual angle, θ_{goal} is the target angle and k_p and k_v are gains. The problem with this is that if the reference trajectory contains sudden changes of angle for any joint the system may be too slow or may overshoot. In the second method, the dynamics of the system is modelled and the theoretical torque needed for each joint is calculated, in real time. It is then possible to anticipate the torque demand. The nominal torque can be fed to the servo system as an extra input and reduces the gain needed and improves performance.

A variant of model reference control which has some advantages is called *algorithmic control*. The term is not used in quite the same way by all authors, but the main principle is that while some joints are forced to follow a reference trajectory others obey servo loops designed to stabilize the system, or optimize some aspect of performance, without regard to the exact path followed. An example is given in the section on reduced order models (see page 101). The term has also been used to describe the version of model reference control in which nominal torques are fed to the servo loops.

It may be noted that finite state control and model reference control represent the extremes of a range of degrees of constraint. In model reference control all spatial variables are completely specified through-out the cycle. In finite state control they are specified only at a few points. As the number of states increases it approaches model reference control; expressed another way, finite state control is a form of sampled model reference control. Algorithmic control is in a sense an intermediate case, since some variables are specified throughout and others are not.

REDUCED ORDER MODELS

Legged robots are exceptionally complicated mechanisms to analyse and control. This is because they usually have many degrees of free-dom. In contrast, a machine such as an internal combustion engine may have many moving parts but these are all connected together and execute a fixed cycle which can be specified by a few parameters. Consequently, attempts to analyse many jointed legged structures (including animals) are often characterized by a search for ways of reducing the number of degrees of freedom to a manageable level. In the worst case, that of a running animal, it is only possible either to model a subsystem such as a single limb (and that not exactly) or to model constrained or simplified motion, for example the com-ponents of force, velocity, etc, in a particular plane. These restrictions apply to most configurations of the legged robot, although with less force as the robot has only one-tenth to one-hundredth as many joints as an animal.

The ways of managing complexity may be summarized as follows:

1) reducing the number of degrees of freedom: analytically, by finding approximations and constraints, and by designing machines with the minimum number of joints,
2) splitting a complex problem into several simpler ones by, for example, separating the control of quantities which do not interact significantly,
3) designing or operating the machine so that the need to control some variables can be avoided, for example providing such a stable base of support that the problem of balance does not arise, or moving so slowly that inertial forces may be neglected.

This section is concerned with the first method. The dynamics of motion confined to a plane is much simpler than that in three dimen-sions. Fortunately, locomotion can often be decomposed into a domin-ant component in the sagittal plane and much smaller components in the frontal and horizontal planes (associated with roll and with yaw, or turning, respectively). Most attention has been paid to the sagittal and frontal planes, since it is in these planes that the problem of balance occurs.

The reduction in complexity is usually achieved by both confining the model to a plane and using a small number of links (which implies that flexible or many-jointed elements such as the vertebrate spine must be drastically simplified). Some of the simpler models which have been found useful will now be described.

SINGLE ACTUATOR MODELS

The simplest case of relevance is that of two linked masses in midair, the masses being typically the body and the leg, or pair of legs moving in unison. The trajectory of the centre of mass of the system is ballistic and is not controllable. What is important is (a) control of body pitch and roll, (b) positioning of the leg for landing. Since there is only one torquer, at the hip, it is not possible to control both of these independently. In the case of the Carnegie-Mellon hopper the body moment of inertia is much greater than that of the leg. No attempt is made to control body attitude in flight; its pitch angle and angular rate are fixed at take-off and assumed to be governed purely by inertia during flight. The hip torquer is used just to position the leg for landing.

This model can also be used to represent the body and tail of a kangaroo-like machine, assuming that the legs are of small moment of inertia or are not moved during flight. In this case, the torquer can react against the inertia of the tail to control the pitch of the body.

When the same machine is on the ground (see Figure 4.5) the hip torquer can be used to control body attitude. The system as a whole is not, in general, stable against falling over, although it can balance

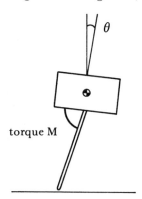

Figure 4.5 *Massive body with light leg and single torquer at hip.*

under some circumstances (Hemami 1973). The Carnegie-Mellon hopper controls its pitch in this way, using a servo of the form:

$$M = k_p(\theta - \theta_{goal}) + k_v\dot{\theta} \qquad (4.2)$$

The Inverted Pendulum

The inverted pendulum (see Figure 4.6) can be used to model the sagittal or frontal plane dynamics of a biped standing on one leg, or of a quadruped standing on two legs and pivoting about the line joining the two supporting feet. It assumes that the foot does not slip and is large enough to sustain the required torque without the machine overturning. The equation of motion in terms of the (as yet unspecified) actuator torque M and the angle θ can be derived in several ways, some of which will be given since although this example is simple it illustrates the use of methods which can be applied to more complex problems.

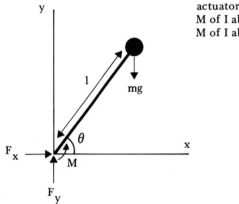

actuator torque is M
M of I about centre of mass is I
M of I about pivot is I_0

Figure 4.6 *Inverted pendulum as a model of a machine standing on one leg.*

One-dimensional (angular) application of Newton's Law

If we regard the pivot as fixed, and the moment of inertia about it is I_0, then the quotation of motion is simply:

$$I_0 \ddot{\theta} = M - mgl \cos \theta \qquad (4.3)$$

This is not applicable to more complex cases as, except for one link, the pivot is not fixed.

Free body with hard constraints

In this method, which can be applied to a multi-link system, each link is treated as a free body, with forces and moments applied at each joint. At each joint there is one actuator torque; the remaining forces and torques (two in the planar case) are due to reaction. These are found by expressing the link coordinates in terms of joint coordinates and differentiating twice. The free body equations for the case described in Figure 4.6 are:

$$m\ddot{x} = F_x \tag{4.4}$$

$$m\ddot{y} = F_y - mg \tag{4.5}$$

$$I\ddot{\theta} = F_x l \sin\theta - F_y l \cos\theta + M \tag{4.6}$$

The constraint equations of the centre of mass are:

$$x = l\cos\theta \tag{4.7}$$

$$y = l\sin\theta \tag{4.8}$$

Differentiating twice:

$$\ddot{x} = -\ddot{\theta}l\sin\theta - \dot{\theta}^2 l\cos\theta \tag{4.9}$$

$$\ddot{y} = \ddot{\theta}l\cos\theta - \dot{\theta}^2 l\sin\theta \tag{4.10}$$

Equations (4.4), (4.5), (4.6), (4.9) and (4.10) can be expressed in matrix form:

$$\begin{bmatrix} m & 0 & 0 & -1 & 0 \\ 0 & m & 0 & 0 & -1 \\ 0 & 0 & I & -l\sin\theta & l\cos\theta \\ 1 & 0 & l\sin\theta & 0 & 0 \\ 0 & 1 & -l\cos\theta & 0 & 0 \end{bmatrix} \begin{bmatrix} \ddot{x} \\ \ddot{y} \\ \ddot{\theta} \\ F_x \\ F_y \end{bmatrix} = \begin{bmatrix} 0 \\ -mg \\ M \\ -\dot{\theta}^2 l\cos\theta \\ -\dot{\theta}^2 l\sin\theta \end{bmatrix} \tag{4.11}$$

Expressing this as:

$$Aq = f(\theta, \dot{\theta}) \tag{4.12}$$

the equations can be solved by matrix inversion:

$$q = A^{-1} f(\theta, \dot{\theta}) \tag{4.13}$$

Note that the equations can be simplified by substitution to eliminate the coordinates x and y, giving:

$$\begin{bmatrix} I & -l\sin\theta & l\cos\theta \\ ml\sin\theta & 1 & 0 \\ -ml\cos\theta & 0 & 1 \end{bmatrix} \begin{bmatrix} \ddot{\theta} \\ F_x \\ F_y \end{bmatrix} = \begin{bmatrix} M \\ -m\dot{\theta}^2 l\cos\theta \\ -m\dot{\theta}^2 l\sin\theta + mg \end{bmatrix} \tag{4.14}$$

This is simple enough to solve symbolically, but in the multi-link case numerical inversion would be necessary. This method gives the forces of reaction F_x and F_y as well as the angular behaviour.

The Lagrangian method

The kinetic energy is:

$$T = \tfrac{1}{2}I_0\dot{\theta}^2 = \tfrac{1}{2}(ml^2 + I)\dot{\theta}^2 \tag{4.15}$$

Lagrange's equation in this case is:

$$\frac{d}{dt}\left(\frac{\partial T}{\partial \dot{\theta}}\right) - \frac{\partial T}{\partial \theta} = F_\theta = M - mgl\cos\theta \tag{4.16}$$

which works out to:

$$(ml^2 + I)\ddot{\theta} = M - mgl\cos\theta \tag{4.17}$$

or alternatively:

$$I_0\ddot{\theta} = M - mgl\cos\theta \tag{4.18}$$

For more than one or two links the expressions become very complicated and automatic procedures are necessary, such as symbolic differentiation by computer. (For references see Hemami 1975.) This approach does not directly yield the forces of constraint.

Other methods

Hemami (1976) describes two other methods: free body with soft constraints and 'superfluous coordinates', which are perhaps of specialized interests.

In all practical cases, for more than two joints numerical or other computer methods are necessary to make use of any of these approaches.

Having found the equation of motion we can try to formulate a control law, that is, an expression for the torque M which will confer postural stability when standing, or will cause a controlled movement from one angle to another. A simple example is a proportional/differential servo. If we redefine the coordinates so that ϕ is the angle from the vertical, the equation of motion is:

$$I_0\ddot{\phi} = mgl\sin\phi - M \tag{4.19}$$

and if the control law is:

$$M = k_p\phi + k\dot{\phi} \tag{4.20}$$

the equation of motion becomes:

$$I_0\ddot{\phi} + k_v\dot{\phi} + k_p\phi - mgl\sin\phi = 0 \tag{4.21}$$

or for small angles,

$$I_0\ddot{\phi} + k_v\dot{\phi} + (k_p - mgl)\phi = 0 \tag{4.22}$$

whose solution is a damped harmonic motion. The design issue amounts to choosing the values of k_p and k_v so that the system will recover from the step or impulse outputs which are likely as a result of disturbances such as a shift of the machine's centre of mass, or subsidence of the soil.

Of course, this control problem can be extended to large angles and other control laws and methods of solving the resulting equation. As with finding the equation of motion with unspecified torque in the first place, solving it for a selected control law usually needs numerical or other computational methods. The design of the control law is a large subject in itself. Further, we may be interested in prescribing a function $\theta(t)$ which we want the angle to follow, rather than simply balancing. It may also be necessary to know the reaction forces at each joint, which may suggest the method to be used.

Development of the Inverted Pendulum

The inverted pendulum model has been developed in various ways. (See, for example, Hemami and Katbab 1982; Bavarian *et al.* 1983.) A variant is shown in Figure 4.7 and can approximate the behaviour of a biped or quadruped under some circumstances, such as swaying with rigid knees.

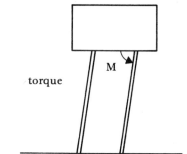

Figure 4.7 *Inverted pendulum with parallel links.*

An alternative approach to the control of a single actuator system with some similarities to an inverted pendulum is illustrated by the frontal plane control system of one of Witt's machines mentioned in Chapter 1 (see Figure 1.6). In a simplified form its frontal plane model is shown in Figure 4.8. This has the features that the machine as a whole is statically stable as long as the centre of mass does not move beyond the space between the legs; but the upper mass is unstable, like an inverted pendulum.

The control problem is to make the lower part of the machine rock from side to side in a stable oscillation using a torque M at the waist joint, reacting against the inertia of the upper mass. (This oscillation produces locomotion as a result of joints and linkages not shown here.) The basic method is to induce a limit cycle using a servo loop of the form shown in Figure 4.9.

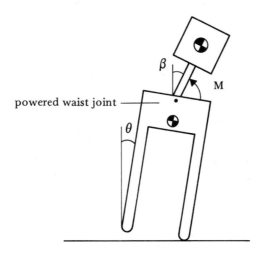

Figure 4.8 *Frontal plane model of walking aid with hydraulic dummy rider.*

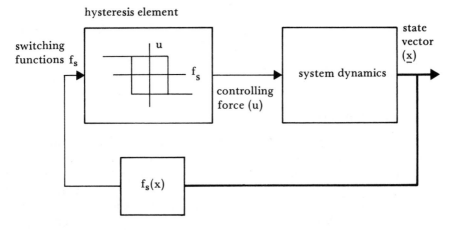

Figure 4.9 *Control system of the dummy rider (after Witt 1977).*

In its simplest form the output of the hysteresis element is a square wave in torque at the waist joint. Switching from one direction to the other occurs when some function of the angle θ crosses a threshold:

$$M = -M_0 \quad \text{for} \quad \theta > \theta_T$$
$$M = M_0 \quad \text{for} \quad \theta < -\theta_T$$
(4.23)

where M_0 is a positive constant. The hysteresis band is $2\theta_T$. The greater it is, the greater are both the period and the amplitude of the oscillation.

However, such a simple control law is inadequate for various reasons, one of which is the inverted-pendulum instability of the upper mass. This requires that β be controlled as well as θ. The control law used is of the form (simplified):

$$M = M_0' - g_1\theta - g_2\beta - g_3\dot{\beta} \qquad (4.24)$$

where M_0' is the output of the hysteresis law shown below (see Figure 4.10) and:

$$f_s = \theta + k_1\dot{\theta} + k_2\beta + k_3\dot{\beta} \qquad (4.25)$$

In these equations, a, M_0, g_1, g_2, g_3, k_1, k_2 and k_3 are constants. (During walking, as opposed to purely planar oscillation, the stride angle and body pitch angle also have to be taken into account.)

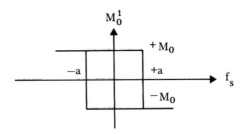

Figure 4.10 *Hysteresis characteristic of the dummy rider control system.*

TWO ACTUATOR MODELS

The number of possible models increases steeply with the number of degrees of freedom. Whereas there are about four single actuator models of interest, there are at least eight significantly different planar models with two controlled joints. Two classes of particular interest are those which model a knee and a hip joint and those which model a hip and an ankle, both obviously being relevant to the single leg support phase of bipedal walking. The two cases differ in that a hip-ankle model has two DOF and two actuators and can stand stably with its joints locked, whereas a hip-knee system has three DOF, only two of which can be completely controlled.

HIP-ANKLE MODELS

The equations of motion have been derived by Hemami and others and are given opposite (see Figure 4.11). They can be found using the methods described in the previous section. The central problem is to formulate laws for M_1 and M_2 to produce either stable standing or part

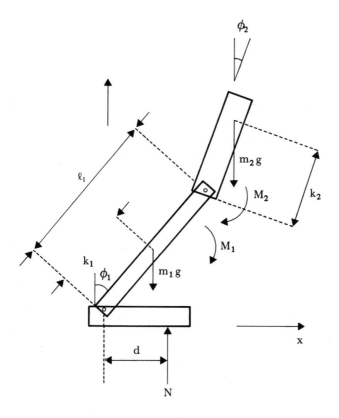

Figure 4.11 *Two mass, two actuator model: the torquers act across the ankle and hip joints. The upper mass represents the body and the lower mass a rigid leg (with knee locked).*

$$(m_1 k_1^2 + m_2 \ell_1^2 + I_1)\ddot{\phi}_1$$
$$+ m_2 \ell_1 k_2 [\ddot{\phi}_2 \cos(\phi_1 - \phi_2) + \dot{\phi}_2^2 \sin(\phi_1 - \phi_2)]$$
$$- (m_1 k_1 + m_2 \ell_1)g \sin \phi_1 = M_1 - M_2 \qquad (4.26)$$

$$(m_2 k_2^2 + I_2)\ddot{\phi}_2$$
$$+ m_2 \ell_1 k_2 [\ddot{\phi}_1 \cos(\phi_2 - \phi_1) + \dot{\phi}_1^2 \sin(\phi_2 - \phi_1)]$$
$$- m_2 k_2 g \sin \phi_2 = M_2 \qquad (4.27)$$

where:

I₁ is the moment of inertia of segment 1 (the leg),

I_1 is the moment of inertia of segment 1 (the leg),

I_2 is the moment of inertia of segment 2 (the body),

m_1 is the mass of the leg,

m_2 is the mass of the body,

ϕ_1 is the angle of the leg from the vertical,

ϕ_2 is the angle of the body from the vertical,

ℓ_1 is the length of the leg,

k_1 is the distance from the ankle joint to the centre of mass of the body,

k_2 is the distance from the knee joint to the centre of mass of the body,

N is the ground reaction force,

d is the distance from the ankle joint to the centre of pressure,

M_1 is the torque of the ankle actuator,

M_2 is the torque of the knee actuator.

---

of the walking cycle. One approach to the second case is to solve the postural stability problem but with the constraint that the body moves from one position to another.

The hip-ankle model has been used by Hemami (1976) to illustrate and compare finite state control and model reference control. It is also a useful example for illustrating some other concepts such as the zero-moment point. The first point which it demonstrates is that for a complex system there is a choice of variables which can be measured. An obvious set is the joint angles, but in some cases it is more convenient to measure other quantities. One set which has been found useful (Vukobratovic and Stepanenko 1972) is N, the ground reaction, and d, the distance of the centre of pressure from the ankle. The centre of pressure is also known as the zero-moment point or ZMP. The equations of motion can be found by taking moments about the ZMP instead of by the Lagrangian or 'hard constraints' method. The real utility of the ZMP is that as long as it is within the area of the foot stability is assured, so it is a useful quantity to control. Of course, controlling it must ultimately be expressed in terms of the joint torques.

A way of applying finite state control to this system is to divide the locomotion cycle into a finite number of phases, the transition from one to the next being triggered by a certain combination of load sensing switches in the sole of the foot. In model reference control the required behaviour of ϕ_1 and ϕ_2 is specified and M_1 and M_2 are calculated from the equations of motion. In algorithmic control d(t) is specified, the ground reaction N(t) is measured continuously and $\phi_1(t)$ is also specified as it would be in standard model reference control. But $\phi_2(t)$, the body pitch angle, is allowed to follow an unspecified trajectory; the equations of motion imply that once ϕ_1, d and N are fixed, then ϕ_2 is determined, since by adding them it is possible to obtain a relationship of the form:

$$f(\phi_1, \dot{\phi}_1, \ddot{\phi}_1, \phi_2, \dot{\phi}_2, \ddot{\phi}_2, d, N) = 0 \qquad (4.28)$$

This happens because specifying ϕ_1, d and N in effect specifies the values of the joint torques; and from this a particular behaviour of ϕ_2 follows.

OTHER BIPED MODELS

An example of finite state control supervising a position servo is presented by Hemami *et al.* (1979). It is applied to a five link model of a biped (see Figure 4.12).

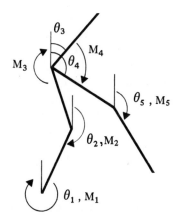

Figure 4.12 *Five link model of a biped:* θ_i *is the angle of segment i and* M_i *is the torque at joint i.*

Each state is not statically stable but is a state of balance maintained by servoing the joints. In no state is there much departure of any segment from the vertical. When in the vicinity of an equilibrium state the system is stabilized by applying a linearized servo in which the torque at each joint is a linear function of all the joint angles and angular velocities. Stability is guaranteed only near the equilibrium points. For stable locomotion the stable regions of adjacent equilibrium points must overlap. It is then possible, having reached equilibrium at one point, to switch the servo to the set of angles specifying the next equilibrium point, and the system will stabilize at this new point. In each state the system is allowed to settle to equilibrium before the new goal equilibrium state is set.

Note that the authors describe this as finite state control supervising model reference control, but the model reference trajectory is not a continuous one. Rather, it is made up of a series of isolated postures each corresponding to a state of the finite state controller.

Kato and co-workers (Ogo *et al.* 1980; Kato *et al.* 1983) have devised a form of locomotion for bipeds which they call quasi-dynamic walking, in which phases of stable movement using ZMP control alternate with moments of 'falling' between one ZMP-controlled phase and the next. This is an example of gait stability. During each single leg support phase the ZMP is found by calculating the combined moment of all the links (seven of them) about the ankle joint, and dividing it by the ground reaction force. The body angle is servo controlled to bring the ZMP back to a nominal position throughout the single leg support phase. The start and end of this phase are signalled by floor pressure sensors in the foot, which also allow measurement of the ground reaction force.

A similar form of gait stability in bipedal walking has been devised by Miyazaki and Arimoto (1983a, b) of Osaka University, by decomposing the movements during the stride into that of the centre of mass of the biped and the relative motion of the leg segments. There are more degrees of freedom than can be completely stabilized, but it is possible, as has been observed earlier, to generate a regular locomotion pattern or gait using an unstable mode. Miyazaki and Arimoto associate the movement of the centre of mass with this mode, using a technique they call the singular perturbation technique. This is possible because the time constant of the unstable mode is larger than that of the other modes, because of the concentration of mass in the upper body.

Beletskii (1979) has analysed a two-stage method of controlling a biped. In the first stage, the joint torques are used to stabilize the system vertically, that is, stabilize its height. When this has been achieved on each stride the system switches to controlling speed and body pitch angle. This is once again a device for stabilizing a system with more degrees of freedom than it has actuators.

NON-REDUCED MODELS

Reduced order models are useful in many circumstances, but there remains a need to handle problems involving many-jointed systems, and in three dimensions without assuming separability into planar motions.

In principle the complete equations of motion can be formulated, using Lagrangian or Newton-Euler methods, and solved (although rarely analytically) but, in practice, automatic methods are needed, so research concentrates on finding efficient methods. There are many different possible situations to consider, for example, whether a linkage forms a closed or an open chain. (A biped alternates between the two: a multipod will usually contain several chains of links, closed through the ground.)

Orin *et al.* (1979) and Walker and Orin (1982) have carried out research in this area. This work was aimed at manipulators, but the methods are also applicable to legged vehicles. It is concerned with formulating and solving the equations of motion of an open chain in a computationally efficient manner. The equations are found by the application of the Newton-Euler equations to successive links in the chain. A coordinate system must be defined for each link, and related to that of the adjacent link by a matrix of direction cosines. A chain of these matrices relates the ith link to a base or absolute coordinate set. It becomes possible to relate the torque at each joint to the position, velocity and acceleration (in translation and rotation) of all the links. Walker and Orin wrote programs based on this technique and compared several variants. Orin and McGhee (1983) apply similar methods to both open and closed chains. Orin and Oh (1980) tackle the inverse problem of finding the torques given the accelerations (which is also discussed in Orin *et al.* 1979), for systems including closed chains. In principle this successive Newton-Euler approach is fairly straightforward; it is only complex computationally.

AN EXAMPLE OF THE ANALYSIS AND CONTROL OF A REAL SYSTEM: THE CARNEGIE-MELLON HOPPER

As described in Chapter 6, Carnegie-Mellon University have been engaged in building a series of highly dynamic legged machines ranging from one-legged hoppers constrained to two-dimensional motion to a four-legged variable gait machine. This section, based on Raibert *et al.* (1983), describes the dynamics of a one-legged hopper which is free to move in three dimensions. It may be regarded in several ways; one is as an instance of a finite state controller (with two states) supervising algorithmic control.

The problem is to stabilize the motion of the machine shown in Figure 4.13. It has one leg, light compared with the body, which can be swung fore and aft and from side to side by hydraulic rams and contains an axial pneumatic cylinder with which it can thrust against the ground hard enough to propel the machine into the air to a height of about half a metre.

The most important achievement of this project has been the demonstration that, at least for one class of machine:

1) locomotion is largely an activity in one plane (the sagittal or longitudinal and vertical plane), with deviations from this plane considered as small perturbations to be controlled separately,

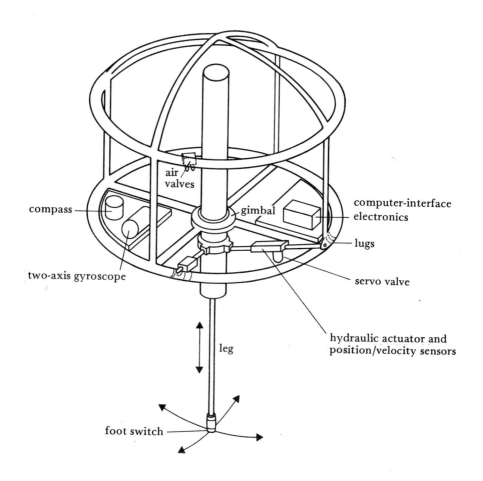

Figure 4.13 *The Carnegie-Mellon three-dimensional hopper [reproduced by permission of M.H. Raibert from* International Journal of Robotics Research — Special Issue on Legged Locomotion *1984, 3 (2)].*

2) the control of locomotion can be divided into three, largely separate, algorithms for motion in the plane, and two more to suppress deviations from the plane.

These algorithms control the following variables:

In plane
1) *height:* hopping height is regulated by measuring the system's vertical energy to determine the correct amount of thrust,

2) *forward velocity:* controlled by selecting position of foot placement and can also be controlled by speed at which leg is swept back during stance,

3) *body pitch:* during stance hip torque is used to erect the body,

Out of plane

4) *body roll:* controlled in the same way as pitch,

5) *spin (yaw):* the foot is placed outside the plane of motion and a torque is generated at the hip. This is translated into a torque about the yaw axis.

The most important of these algorithms are now described.

The machine hops by bouncing on its springy leg and alternates between periods of support and periods of flight. In flight, motion of the machine as a whole is ballistic. During the support, or 'stance', phase it behaves as a two-link inverted pendulum. The analysis of flight uses one set of equations and that of stance a second set. A third part of the analysis deals with the instantaneous transition from flight to support and vice versa, at which points the conservation of momentum is used to determine the change in velocities due to the transition.

Control Decomposition

Observation of running animals suggested that most of their motion occurs in the sagittal plane; the swinging of the legs and the oscillations of the body are largely confined to this plane, with relatively small movements used to correct deviations and to steer.

The control of motion of a one-legged hopper in the plane can be done by the first three algorithms listed earlier. Extra-planar control must deal with three sources of error: roll (falling over sideways), yaw (spinning about the vertical axis) and sideways translation. Roll and yaw are controlled by the fourth and fifth algorithms. Lateral errors are not controlled as such; rather, since the machine is more or less rotationally symmetrical about the vertical axis, its direction of motion can be redefined at each step. This implies that the machine in its present form cannot make sudden changes of direction while moving at speed; unlike a running animal it cannot lean into a bend.

The Mathematical Model

The hopper is modelled as a rigid body connected to a leg by a universal joint at the hip. The two DOF of this joint are driven by torque actuators. The leg is modelled as an axial position actuator in series with a spring. (The springiness is provided by the air in the actuator.) Both body and leg have a significant mass and moment of inertia, and the ground can be springy, but this resilience is often negligible. This model, in its two-dimensional form restricted to motion in the plane of travel, is shown in Figure 4.14. The leg has mass M_1 and the moment of inertia I_1, and the body has mass M_2 and moment of inertia I_2. The centre of mass of the leg is located a distance r_1 from its lower tip, and that of the body is a distance r_2 from the hip joint.

The hip torquer is used in an angular position servo loop described by the equation:

$$\tau(t) = K_p(\theta_1 - \theta_{1d}) + K_v(\dot{\theta}_1) \qquad (4.29)$$

where:

τ is the torque generated,
θ_1 is the current leg angle (measured from the vertical),
θ_{1d} is the desired leg angle,
K_p and K_v are feedback gains.

In other words, the hip torque is proportional to the leg angle error and to the leg angular velocity. This control law applies on the ground and in flight, but with different values of K_p and K_v.

The leg is modelled as a position actuator between the body and a spring. The stroke, χ, is limited by mechanical stops. For the purpose of analysis the stop which limits extension is modelled by increasing the stiffness of the spring when it is in tension beyond its value when in compression. This stiffening is accompanied by damping to eliminate the vibrations which would otherwise occur when the spring is fully extended.

When the machine is on the ground its vertical behaviour can be modelled by a spring-mass oscillator with natural frequency:

$$W_n = \sqrt{\frac{K_2}{M_2}} \qquad (4.30)$$

where K_2 is the stiffness of that leg spring. Each stance interval has duration:

$$T_{ST} = \frac{\pi}{W_n} = \pi\sqrt{\frac{M_2}{K_2}} \qquad (4.31)$$

During flight the machine follows an arc whose duration is:

$$T_{FL} = \sqrt{\frac{8H_0}{g}} \qquad (4.32)$$

where H_0 is the height of the hop (that is, the change in height of the centre of gravity of the whole machine).

So a full hopping cycle has the period:

$$T = \pi\sqrt{\frac{M_2}{K_2}} + \sqrt{\frac{8H_0}{g}} \qquad (4.33)$$

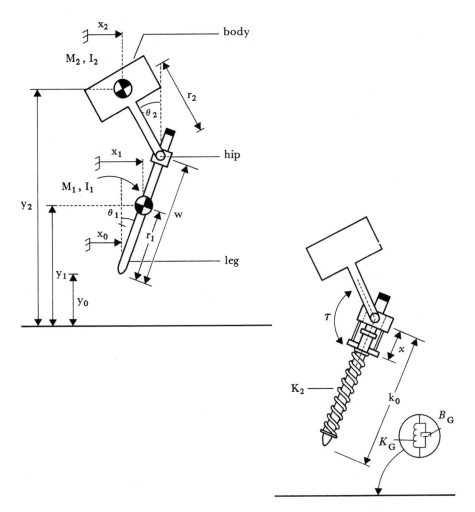

Figure 4.14 *Simulation model of the planar hopping machine*
[reproduced by permission of M.H. Raibert from
IEEE Trans. on Systems, Man and Cybernetics *1984,* **14** *(3)].*

It is important to know how much the leg must shorten during the stance phase. The maximum compression ΔW of the leg spring during stance is:

$$\Delta W = \frac{2M_2 g}{K_2} + \sqrt{\frac{M_2^2 g^2}{K_2^2} + \frac{2M_2 g H_0}{K_2}} \qquad (4.34)$$

It is possible to derive the equations of motion of this model using d'Alembert's principle. They are a coupled set of non-linear differential equations. These can be expressed in terms of the five state variables, θ_1, θ_2, x_0, y_0 and W. These equations, which are rather complicated, are given in Raibert *et al.* (1982). They have been solved by numerical integration to simulate the behaviour of the system under various conditions.

Vertical Control

The fundamental method is to set up a stable oscillation in which on each hop the energy needed to reach the desired height is calculated and used to control the amount of air injected into the leg actuator. This procedure can be divided into the following steps:

1) calculate what the kinetic energy of the system would be just after lift-off if no air were added, taking into account the energy losses at touch-down and lift-off,
2) calculate what kinetic energy E_H is needed to make it reach a specified height H,
3) the difference between these is:

$$\Delta E_H = E_H - E_{LO}. \tag{4.35}$$

It must be supplied by injecting air at the bottom of the cycle.

It can be assumed that the only major energy losses are at lift-off and touch-down. The energy loss at touch-down is just the vertical kinetic energy of the leg.

This algorithm leaves certain characteristics of vertical control unspecified. One of these is the time at which the leg is retracted, as it must be at some point in the cycle. This time can be chosen to optimize performance according to some criterion, for example:

1) shortening at lift-off optimizes ground clearance,
2) shortening at touch-down minimizes ground impact forces.

Velocity and Pitch Control

Balance, the control of horizontal velocity and the control of body pitch angle are all related. (Balance follows from horizontal velocity control. If the velocity is controlled, this implies that balance is being achieved.) They are all effected using the same actuator, the hip torquer. This single actuator can control both velocity and attitude because it operates in two different regimes alternately: during stance friction between the foot and the ground allows hip torque to rotate the body; while in flight the inertia of the body allows the angle of the leg to be controlled. Several algorithms have been tested, but only the simplest will be discussed here.

Velocity

Forward velocity is determined at each hop by the angle of the leg at touch-down. The strategy used is to maintain balance by keeping the average tipping moment over each stride approximately equal to zero. If there were no errors this would imply that for constant velocity the leg angle at touch-down should be equal in magnitude to that at lift-off:

$$\theta_{1TD} = -\theta_{1LO} \qquad (4.36)$$

The analysis begins by defining the *CG-print*, which is the locus of points over which the centre of gravity of the body travels during the stance period. If the duration of stance is T_{ST} and the horizontal velocity of the body is $\dot{x}_2$, then the length of the CG-print is:

$$\Delta x = \dot{x}_2 T_{ST} \qquad (4.37)$$

To minimize tipping, the foot should be placed at the centre of the CG-print:

$$x_0 = x_{2TD} + \Delta x/2 \qquad (4.38)$$

If the foot is placed forward of this, the hopper will tend to tip backwards, which will slow it down, and vice versa. A linear function of velocity error is used to deviate the foot placement from the centre of the CG-print:

$$x_{ERR} = K(\dot{x}_2 - x_{2,D}) \qquad (4.39)$$

where $x_{2,D}$ is the desired value for $\dot{x}_2$ and K is a feedback gain. Therefore, the desired foot position to correct any velocity error is:

$$x_0 = x_2 + \frac{\Delta x}{2} + K(\dot{x}_2 - \dot{x}_{2,D}) \qquad (4.40)$$

and the corresponding touch-down leg angle is:

$$\theta_{2TD} = \arc \sin \left[\frac{\dot{x}_2 T_{ST} + 2K(\dot{x}_2 - \dot{x}_{2D})}{2W} \right] \qquad (4.41)$$

So velocity can be controlled by selecting this angle. The hip torque servo system is used during flight, reacting against the body's inertia, to move the leg to the right angle before touch-down.

Body Pitch Angle

During stance the same hip servo, but with different gains, is used to rotate the body to the desired attitude, this time reacting against the ground by way of the leg.

Multipod Control

The basic design problem for the control of a multipod has the four steps listed on page 98 but the emphasis is different. In particular, specifying the required behaviour is complex. The simplest approach is to begin with a desired path (in position and orientation) of the body and from this calculate the leg movements needed to produce it. This approach is indeed useful, but of course it is not possible to specify a completely arbitrary trajectory since the legs have only a finite range of movement and a limited range of stable support patterns. A more complex approach is to treat the body trajectory not just as a completely specified path, for example a straight line between starting point and destination, but to generate a nominal trajectory which optimizes some aspect of performance. In other words, the details of the path may be chosen to make things easier for the legs.

Whichever approach is adopted, the problem tends to decompose into a number of areas:

1) specifying a body trajectory,
2) for a segment of a body trajectory during which a particular set of feet is on the ground, finding a set of joint angle trajectories which will produce the body path,
3) finding the joint torques to produce these angle trajectories,
4) finding, whenever a leg reaches the end of its stride, a new foot placement for it.

As implied previously, these goals may interact.

The terms gait selection and gait implementation are often used in connection with the multipod control problem, but somewhat variably. Gait certainly involves area 4 of the list and is sometimes considered to involve the other areas. Gait selection might be regarded as the prior decision to stick to, say, the alternating tripod gait; the control problem would then be described as gait implementation.

It should be noted, in connection with areas 2 and 3 in particular, that there may be several possible sets of joint angle and torque patterns which will produce a selected body trajectory. The issue then arises of controlling additional aspects of performance. Typical aspects are: compliance ('springy' suspension), load distribution among legs, avoiding opposing forces in different legs and optimizing power consumption or acceleration.

This section reviews some approaches to these problems, starting with gait generation. The simplest case is that of walking in a straight line on a plane, obstacle-free surface. As explained in Chapter 2, a gait can be chosen to optimize some property such as speed or stability. The gaits most intensively studied have been wave gaits, which optimize longitudinal stability. This section shows how the times of foot placement and lift-off may be chosen to implement a wave gait.

WAVE GAIT IMPLEMENTATION FOR WALKING IN A STRAIGHT LINE

Waves of stepping sweep from back to front on each side. The timing of all events is taken to be relative to touch-down of the left front leg. For each leg we define a relative phase ϕ_i in the range 0 to 1, by which it lags the reference leg. Each leg on the right side lags the corresponding left leg by a phase of 0.5. An important parameter of the gait is β, the duty factor, which is the fraction of the cycle for which a leg is on the ground. It is assumed to be the same for all legs and for a wave gait $0.5 \leqslant \beta < 1$. The relative phases are shown in Figure 4.15.

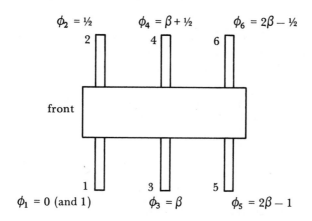

Figure 4.15 *Definition of leg numbers and relative phases for a hexapod executing a wave gait.*

They are derived from the equation proved optimal for longitudinal stability by Bessonov and Umnov (1973):

$$\phi_3 = \beta, \quad \phi_5 = 2\beta - 1, \quad \beta \geqslant 0.5 \tag{4.42}$$

An example of an event sequence for one cycle, where $\beta = 0.75$, and ignoring the events which overlap from the previous cycle, is shown as a gait diagram in Figure 4.16. It can be seen that the front left and back right legs move in unison, as do the front right and back left.

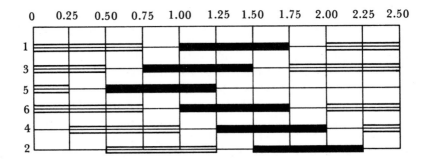

Figure 4.16 *Gait diagram of hexapod wave gait.*

It is convenient to relate relative phase to absolute time and distance. We define:

stride length, λ, is the distance the body moves in one cycle.
stroke, S, is the distance it moves when a leg remains on the ground,
period, T, is the time required for one cycle,
forward speed, u,
kinematic cycle phase, ϕ, is the distance the body has moved since the last placement of leg 1, normalized to the stride length.

The following relationships hold:

$$\text{stroke} \quad S = \beta\lambda \tag{4.43}$$
$$\text{speed} \quad u = \lambda/T \tag{4.44}$$

and if t is time measured from the start of a cycle:

$$\phi = t/T \quad \text{or} \quad \phi = \int \frac{1}{T}\, dt \tag{4.45}$$

if u is not constant.

The time of touch-down for each leg during forward walking is given by the moment when $\phi = \phi_i$. Lift-off occurs at:

$$\phi = [\phi_i + \beta]_{\text{mod } 1} \tag{4.46}$$

A way of dealing with backward walking is to give a negative sign, and change these two equations to:

$$\text{touch-down} \quad \phi = [\phi_i + \beta]_{\text{mod } 1} \quad T < 0$$
$$\text{lift-off} \quad \phi = \phi_i \quad\quad\quad\quad T < 0 \tag{4.47}$$

respectively.

GAIT EXECUTION IN MORE COMPLEX SITUATIONS

The analysis and selection of gait can be extended in several directions beyond straight line walking, even without taking dynamics into account. The first of these to be described here is the control of leg movement so as to implement a regular gait but still allow the vehicle to turn, and then the selection of foot placement when forbidden zones such as obstacles must be avoided.

Regular Gait Allowing Turning

The simplest way of steering a multipod is to provide a steering mechanism separate from gait generation. This has been done or proposed in several ways. The Komatsu underwater octopod (see Chapter 6 and Ishino *et al.* 1983) uses an alternating tetrapod gait. Each set of four legs is attached to a frame which moves as a whole. The two frames can rotate relative to each other about a vertical axis as shown in Figure 4.17.

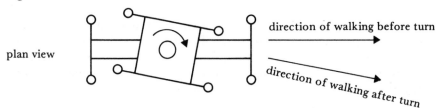

Figure 4.17 *The steering mechanism of ReCUS: the square represents the inner frame, which can slide along the outer frame and also rotate about the central pivot. The inner and outer frames each have four legs at the corners.*

To turn, the machine stands on one set of legs and rotates the frame bearing the raised legs.

A method for hexapods using the alternating tripod gait is shown in Figure 4.18. It is intended for machines where the propulsion stroke is not a rotation about the hip but occurs when the lower part of the leg swings in a vertical plane about the knee. If an extra vertical pivot is fitted at the knee of the front and rear legs, then the machine can pivot about the middle foot on the opposite side.

If the robot has three more or less orthogonal controlled DOF per leg, as is true of the more advanced hexapods, steering and other motions can be achieved by calculating the required trajectory of each foot relative to the body and applying this to the leg servos. If the body path is circular, then the position of the foot relative to the body will also be a circle, centred on the turning point. Bessonov and Umnov (1980) have discussed control schemes for achieving this electronically, using leg position signals to control the speed of the motors without

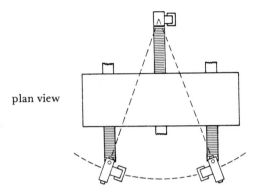

plan view

a left turn is illustrated
supporting legs are shaded

Figure 4.18 *A steering mechanism for a hexapod: the plane in which
each end foot swings can be rotated about a vertical axis through the knee;
this allows the machine to pivot about the middle on the opposite side.*

computation, which is possible because a circular arc is a simple
function to generate. A more general approach is described in Orin
(1982) in which the body-relative foot velocities are calculated as a
function of body velocity, rate of turn and present coordinates of the
foot. For the geometry of Figure 4.19 the result is:

$$\begin{bmatrix} \dot{x}_i \\ \dot{y}_i \end{bmatrix} = -\begin{bmatrix} u \\ v \end{bmatrix} + \begin{bmatrix} \dot{\theta} y_i \\ -\dot{\theta} x_i \end{bmatrix} \qquad (4.48)$$

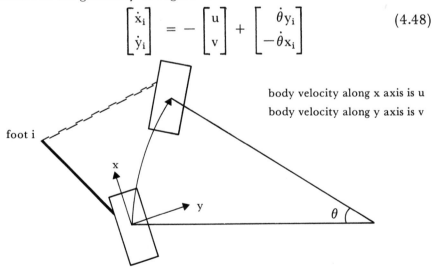

body velocity along x axis is u

body velocity along y axis is v

foot i

Figure 4.19 *Geometry for calculation of components of foot velocity
during a turn.*

This velocity equation applies to a foot during its support phase, the
timing of which is still governed by the gait cycle. But unlike the

straight line case an additional issue here is to put down each foot at the start of its support phase. In the same paper by Orin the method chosen is to select the foothold so that the leg will be in a reference position in mid-stride. The reference position is that which would be occupied in straight line walking. The equation given above, together with other simple relationships, can be used to calculate where to put down the foot so that, given the body's present trajectory, the foot will pass through the reference position at mid-stride.

A final consideration is how to get the foot from its position at the end of one support phase to its touch-down position at the start of the next. Orin calculates a constant horizontal velocity which uses all the time available until touch-down, and a half sine wave to represent the vertical profile of foot movement during the transfer.

Free Gaits

A gait is usually a regular cycle but need not be. If the ground is very uneven, then a regular gait may not be possible. In this case, the next foot to be placed is selected not by its place in a fixed sequence but by some optimality criterion. This is the procedure used by human rock climbers; the limb to be moved next is usually the one which can contribute the least further movement on its present hold, or it may be the one which can secure the best new hold. (Choosing the 'best' involves a combination of position and security.)

Such gaits have been termed *free gaits*. They have been studied by Kugushev and Jaroshevski (1975) and by McGhee and Iswandhi (1979). McGhee and Iswandhi considered a multipod walking, slowly enough for static stability to be assumed, on level ground containing forbidden zones representing obstacles. It is assumed that there is some way of detecting forbidden zones, certainly within the reach of the vehicle's legs at a particular moment and preferably some way ahead. The algorithm is essentially the rock climber's algorithm. As the machine moves steadily forwards the control program continuously tests to see whether it is desirable to lower a new leg and raise a supporting one.

The algorithm makes use of the notion of the *kinetic margin* F_{ij} of a leg j when placed on a foothold i. It is, in effect, the distance the machine can move without becoming unstable before that leg reaches the end of its motion. If the kinetic margins of two legs are compared, the one with the greater is the better bet. Such comparisons are used both for deciding which of the supporting legs should be raised first (the one with the smallest kinetic margin) and for determining when it becomes better to replace this leg by a new one, and with which new leg it should be replaced. Whenever a new leg is placed, the one with the largest kinetic margin is chosen. To identify this leg and decide where to place it, it is necessary to examine all the possible valid footholds that can be reached by every currently raised leg. This is done by dividing the ground into squares labelled accessible or forbidden. All accessible cells within the range of movement of each leg are examined and the kinetic margin of the leg calculated in each case.

The algorithm keeps the minimum number of legs on the ground. This is usually three as might be expected. It will lower a new one if static instability is imminent. Thus, there are two reasons for lowering a leg: instability, and a leg running out of reach. A simplified flow chart is shown in Figure 4.20. Note that it is possible for the program to get stuck in a loop.

Hirose (1984) has devised an approach which combines the benefits of both a regular gait and a free gait. The problems with a free gait are that it is slow and energetically inefficient and, as has just been seen, can lead to a situation of 'deadlock' in which further movement in the desired direction is impossible. Hirose's algorithm is based on the idea of conforming to a regular optimal gait as nearly as possible. It does this by restricting the area of search for the new foothold according to a series of principles which together avoid deadlock and converge to the standard gait.

The first principle restricts the forward swing of a leg so that the base of support for the next step is not reduced too far. The second and third principles aim at maintaining the standard quadruped crawl by restricting the foothold search area so that the correct leg can be moved on the next step while preserving stability. The application of these principles to the reachable area of the leg results in a search area which must be searched until a point in it is found which also lies within a safe (non-obstacle) region, as shown in Figure 4.21. The order of search is determined by *phase-preserved positioning*. An ideal foot position is calculated which would conform perfectly to the standard gait. The search area is then scanned starting at the end nearest this ideal point. The search will then find the nearest point which is reachable, safe and conforms to the three principles.

An interesting discovery was that (in simulation) this algorithm resulted in the automatic orderly crossing of a river-like obstacle as wide as the leg stroke, with a resumption of the standard gait afterwards.

ACTIVE COMPLIANCE

The problem of distributing a multipod's weight and inertia load among its legs, and of producing a smooth ride, requires a degree of springiness in the legs. This may be achieved by genuine springs (or the elasticity of pneumatic actuators) or artificially by implementing a servo control system which imitates the behaviour of a spring. This approach has been tried on the Ohio State University hexapod with

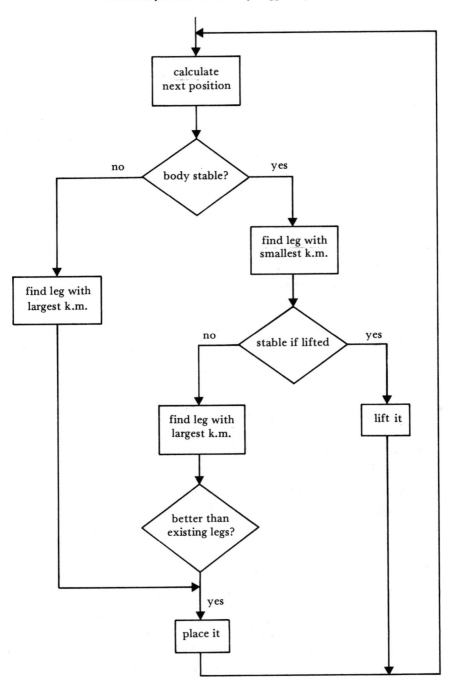

Figure 4.20 *Free gait algorithm.*

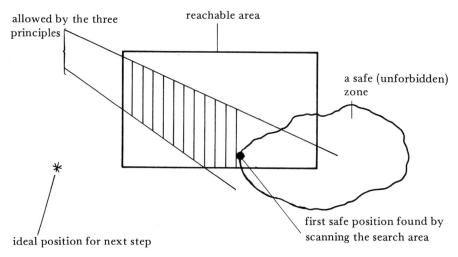

allowed by the three principles

reachable area

a safe (unforbidden) zone

ideal position for next step

first safe position found by scanning the search area

Figure 4.21 *Search area for a constrained free gait (phase preserved positioning).*

some success (Klein *et al.* 1983). The basic method is to cause the height of each leg to obey a law of the form:

$$\dot{z} = \dot{z}_D + k_p(z_D - z) + k_f(F_D - F) \qquad (4.49)$$

where:

z and ż are the actual vertical position and velocity,
z_D and $\dot{z}_D$ are the desired vertical position and velocity,
F and F_D are the actual and desired force,
k_p and k_f are gains.

Considering the case of constant desired position and desired force and substituting for the inertial component mż of the actual force (m is the mass), the equation can be transformed into:

$$A\ddot{z} + B\dot{z} + Cz + D = 0 \qquad (4.50)$$

where A, B, C and D are constants. This is the equation of a damped harmonic motion, so the leg behaves as if it contained a spring and a damper.

The method assumes that the leg servo system is fast and accurate enough for the actual velocity to follow the commanded velocity as specified by the control law. This could be a significant problem if a stiff 'effective spring' is required, as might be the case for high speed locomotion.

This active compliance method was judged effective, although low pass filtering of the force signal had to be added to eliminate modes of oscillation which were discovered when all legs were active.

CONTROL OF BODY ORIENTATION AND HEIGHT

If a machine stands on three or four legs it is clearly possible to control body height by a collective movement, and body attitude by a differential movement, of the legs. A sensor, usually a combination of gyroscopes or gyroscopes and pendulums, is needed to measure the pitch and roll angles. Height, or perpendicular distance from the surface, can be calculated from the leg geometry. Hirose (1984) describes an experiment in which the body of a quadruped is kept level by sensing tilt with a pendulum sensor. When the body tilt causes the pendulum to touch a contact a command is generated to extend or retract the legs in such a sense that the body is levelled.

One of the simplest methods of keeping a more or less constant attitude and average height is shown in Figure 4.22, which shows a side view of an octopod with four side-by-side pairs of legs climbing stairs.

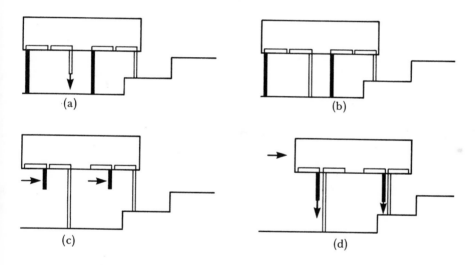

Figure 4.22 *Stair climbing with constant body attitude.*

In Figure 4.22a the vehicle stands on the shaded legs, at the end of a forward movement; the other legs (unshaded) are being lowered. The front pair has touched the ground and is stopped, but the rear pair continues to descend until it touches the ground (see Figure 4.22b). In Figure 4.22c the unshaded legs have all been extended an equal amount further, raising the body; the shaded legs have been retracted. In Figure 4.22d the body has moved forward again and the shaded legs have been advanced. They can now be lowered, repeating the cycle.

Such a system must still incorporate body attitude sensing if it is to avoid a cumulative build up of errors, unless it always walks on surfaces which are piecewise level.

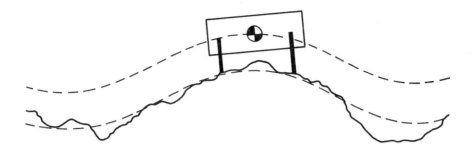

Figure 4.23 *Body attitude when walking on a rough, undulating surface.*

A more complicated problem is shown in Figure 4.23. Unlike the previous example, we wish to avoid steps in the body's trajectory, but it is impractical for the path to be a straight line; rather, it must be a smoothed version of the surface profile. (It may or may not be feasible to keep the body attitude constant.)

First, note that if the leg joint velocity could be continuously controlled, the stair climbing method shown in Figure 4.22 could be applied while the body moved in a straight line parallel with the average slope of the stairs; this would imply that, among other things, at stage b in Figure 4.22 when the front pair of legs touched the ground they would not stop completely but would continue to extend at the (slower) rate needed to match the body's steady rise in height. Also, at the same time they would have to start moving backwards relative to the body so as not to be dragged along by its forward movement.

This suggests that the solution to the problem described in Figure 4.23 is to:

1) measure the ground profile,
2) calculate a smoothed version to use as the body trajectory,
3) use the algorithm for stair climbing supplemented by corrective motions to eliminate cumulative errors in attitude and mean body height.

The problem is also discussed by Bessonov and Umnov (1983) and Klein *et al.* (1983).

Chapter 5
Computing: hardware and software requirements

In this chapter the requirements of a computer control system for a walking robot are discussed. Basic concepts are dealt with rather than specifics such as interfacing. The first two of four sections deal with hardware requirements and the last two with software.

Computer Architectures

A general model of a robot control system is shown in Figure 5.1.

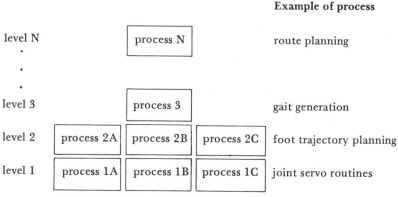

Figure 5.1 *A general model of a robot control system, showing some examples of the processes controlled by such a system.*

Its main characteristics are that it is made up of processes, some of which may run simultaneously, and that these processes can be divided into levels. Some examples of processes controlled by such a system are shown in the right-hand column of the figure. A typical process is the execution of a servo control algorithm for a single joint of a robot leg (or arm). Alternatively, a single process might control all the joints of the leg. It is not intended that this description be a rigorous definition of a process, but it is usually thought of as the continuous execution of some program in real time, so that it cannot be arbitrarily stopped and started, or run at an arbitrary speed.

It may be assumed that most current and future generation robots will be controlled by digital computers, certainly at the lower levels concerned with gait generation and servo control. Control at the very highest level, that of selecting and initiating the task for the robot, will always be exercised by a human user.

Robots vary in the level at which the division of control between machine and user occurs. Also, the division may be a three way one, between the robot itself, an intermediate computer (which may also control other machines) and a human user who specifies the aims of the whole system. This arrangement is often found in mobile laboratory robots, where a small on-board computer interrogates sensors and drives the wheels, while a larger fixed computer handles navigation and other complex tasks.

The model shown in Figure 5.1 can represent the whole system including the user, or it may stand for just that part handled by the computers. In either case, the robot designer must decide how to provide the computing resources needed to run the required processes. The following questions must be answered:

1) Is one computer or several to be used, and if several, how many?
2) How are the control tasks to be allocated among the processors?
3) How are the processors to communicate with each other?
4) What is to be the nature of the user's interface?

These questions of architecture cannot be isolated from the questions of what language(s) and operation system(s) or other run-time environments are to be provided. Such decisions as whether to use a single computer or to have several smaller processors each running a single process, must take into account the availability of suitable multitasking software. These issues are of course not unique to legged vehicles but arise in real-time control generally, and there is an extensive literature on real-time languages, inter-processor communication and so on (eg Young 1982).

The following sections report some architectures which have been used for legged robots.

SINGLE PROCESSOR

Many robots use a single computer executing several processes in parallel. In the example shown in Figure 5.2, which is a simplified version of the Ohio State University (OSU) hexapod control system (Orin 1982; Klein *et al.* 1983), the main processes are motion planning and servo control. Various models of the PDP-11 computer have been used. It should be noted that it is not always necessary to have an interrupt-driven multi-task system. Robots have been controlled by a single computer running a single task which implements a finite state machine (Todd 1984).

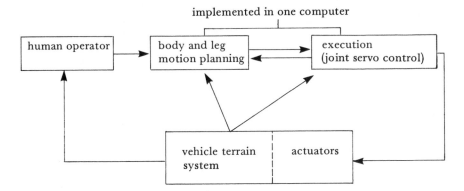

Figure 5.2 *A control system using a single computer.*

SINGLE LAYER NETWORK

One of the simplest multiprocessor architectures is shown in Figure 5.3a.

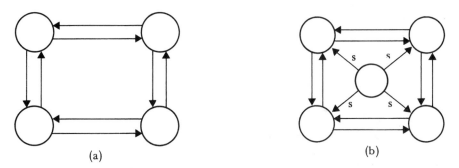

Figure 5.3 *Single layer networks: (a) four equal processors with interlocking signals; (b) the same with central synchronization.*

Here all processors are of essentially equal status and each runs one process, controlling, for example, one leg. Each may run to a large extent independently of the others, but some synchronizing or interlocking signals do need to pass between them. An *interlocking signal* is one emitted by one processor to delay a second until the first has completed some necessary action. An example might be that of a signal to prevent a leg being raised until all the others had been lowered. A system linked only by interlock signals would run at a speed determined by the times taken to complete the various actions and which would not necessarily be constant or controllable. A *synchronizing signal* is one which causes a processor to execute some reference phase of its activity at a specified time or with a specified period. Synchronizing signals for all the processors may originate in a central source, as in Figure 5.3b.

The processors need not in principle be programmable digital computers. In the Phony Pony each was a finite state machine made of a few flip-flops.

The coordination of arthropod walking has been modelled by networks of this kind in which each leg has a processor in the form of a ganglion or a small group of nerve cells. Each ganglion may constitute an oscillator capable of generating a stepping cycle for its leg, but the timing (and other parameters) of whose oscillation is affected by external inputs. In the case of an insect, signals from the rear legs, in this model (see Figure 5.4), trigger the middle legs after a delay, and the middle legs trigger the front ones. Consequently, waves of stepping sweep from back to front. Such a model is not supposed to be complete; there is also supervision by the brain.

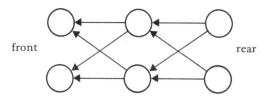

front   rear

Figure 5.4 *Model of insect coordination.*

SINGLE MASTER WITH ONE SLAVE PER LEG (OR JOINT)

If in the previous architecture the central synchronizer is given a more complex, coordinating role and the communication between the other nodes is reduced to a minor role or eliminated altogether, the architecture shown in Figure 5.5a is arrived at. The task of the lower level processors is to execute some function such as joint servo control

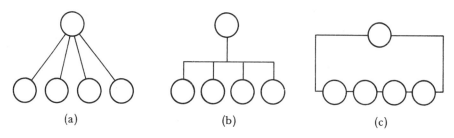

(a)   (b)   (c)

Figure 5.5 *Architectures having a single master with several slaves:*
(a) individual connection between master and each slave;
(b) bus connection; (c) ring connection.

(or the control of a whole leg's tip position) which is independent of direct feedback from other legs; the supervising processor takes sensory information from the whole system and issues commanded target angles, torques or rates to the joint or leg controllers. This architecture is found in the Odex 1 (Russell 1983) in the variant shown in Figure 5.4b, where the connection is by a bus, under the control of the master processor.

If a ring is used (see Figure 5.5c), the architecture is distinguished from that shown in Figure 5.3a only because one processor dominates the others.

PYRAMID

The previous architecture is a two level hierarchy; more levels are of course possible. The branching factor need not be greater than one at all levels. An example is shown in Figure 5.6.

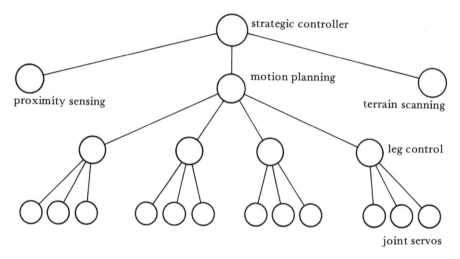

Figure 5.6 *Pyramid or tree structure.*

An example of a multiprocessor

As an exercise in the design of multiprocessors for legged robots, the group at Ohio State University have split up the functions of Figure 5.2 in various ways and distributed them among a group of five processors (Klein and Wahawisan 1982). The basic configuration is shown in Figure 5.7. It consists of five PDP-11/03s, each connected to all others by parallel links. Removal of selected links can reconfigure the network as a ring, star or tree. In the experiments carried out at OSU, the connection with the hexapod was made via a PDP-11/45, but in principle each 11/03 could have its own connection to part of the robot, for example a pair of legs.

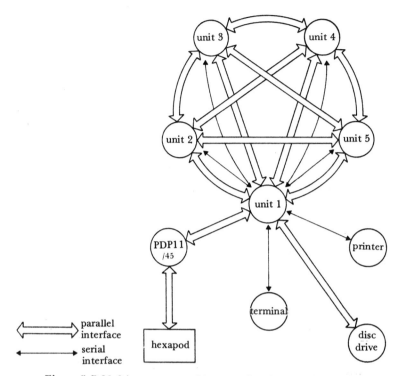

Figure 5.7 *Multiprocessor architecture for the OSU hexapod.*

Among the configurations which have been tried are a master and three slaves, in which the master carries out body motion planning and generates joint rate commands for a pair of legs. A second configuration, whose purpose is fault-tolerance, uses a ring of processors among which the control tasks are shared. The ring can be shortened to isolate a faulty processor — this is a system of the kind shown in Figure 5.5c. All the processors are programmed in Pascal. The master runs it under RT-11; the slaves or satellites run Pascal code compiled on and loaded from the master, without an operating system. Interprocessor communication is handled by assembly language macros.

Sensing Methods

The purpose of this section is to describe the sensing methods which have been found useful for legged robots. The sensory problem may be summarized thus: Is the following known?

1) the topography and physical properties of the terrain,
2) the translational and rotational position, velocity and acceleration of the robot as a whole, either absolutely or relative to the terrain,

3) the disposition of its movable parts, principally its legs,
4) the forces on its rigid segments and the torques about its joints.

In practice it is possible to sense or infer a set of quantities that is partially redundant and usually incomplete, from which the four listed kinds of information need be determined. There are often alternative ways of specifying them; such alternatives are familiar. For example, in navigation a continuous record of heading together with distance travelled from a known point is equivalent to latitude and longitude, and to three bearings to known points. Another example is the relationships of differentiation and integration between position, velocity and acceleration. Making certain assumptions, any one of these can be used to determine the others.

When alternative sources of information exist, each usually has weaknesses which imply that some kind of combination of sources is required. In cases where the 'weakness' can be expressed in terms such as signal-to-noise ratio and frequency of measurement there are mathematical ways of doing this, for example Kalman filtering, but the sensing problems of legged robots tend to be too complicated for such techniques to be sufficient on their own. However, this chapter is not concerned with the problems of combining different sensor signals, but simply with individual sensing techniques. A list of quantities of interest is given below:

1) joint angle and angular rate,
2) force sensing,
3) touch sensing,
4) proximity sensing and short distance range measurement,
5) body orientation and angular rates,
6) body position, speed and acceleration,
7) terrain scanning,
8) soil properties.

Many of these are not confined to legged robots but also apply to other vehicles.

JOINT ANGLE AND ANGULAR RATE

Extreme precision is not usually necessary. Unlike a target position for the end effector of an industrial robot, a target foothold or a target body movement is not usually very precisely defined. Most walking machines have used potentiometers for joint angle feedback. A potentiometer can be very cheap compared with a digital encoder and has the, perhaps minor, advantage of needing only three wires. Nevertheless, the trend for all sensors to be digital will probably eventually affect joint angle measurement.

Angular rate, if differentiation of the joint angle signal is not accurate enough, can be measured using a separate tachometer. For electric actuation this can be fitted to the motor before the reduction gearing, therefore reducing the effect of backlash.

FORCE SENSING

It may be useful to know the force on almost any part of a walking robot, but the forces which have been judged most important so far are those in the leg near the foot.

For a very small foot, as has been usual with multipods in the past, what is usually measured is the axial force and two orthogonal components of the transverse or bending force on the most distal segment of the leg. In a simple case (see Figure 5.8a) some of these axes may align conveniently with body, ground and gravity axes; in a more general case this is not so. To take the axial component, for example, it may be necessary to distinguish between the axial, vertical, local surface-normal and vehicle-averaged surface-normal directions (see Figure 5.8b).

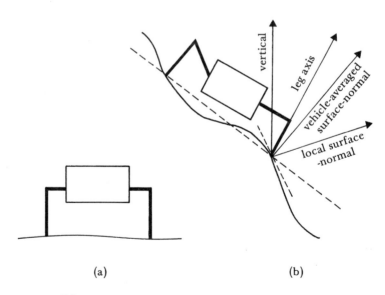

(a)   (b)

Figure 5.8 *Definition of axes for specifying forces:*
(a) cross section of vehicle on level ground;
(b) vehicle on rough sloping ground.

To illustrate the uses of force sensing, consider a machine whose distal segments are more or less normal to the surface. The differential forces in the surface plane are then an indication of whether the legs are acting against each other. The force on a foot in the direction of the propulsion stroke is a measure of whether it is gripping firmly or slipping. The differential axial or surface-normal force reflects any imbalance in the load distribution among the legs. The ratio of the collective axial force to the collective surface-parallel force is a measure of slope. It is also possible to determine whether the vehicle is in danger of overturning. If the robot has extended feet it may be desirable to measure forces at several points along the feet.

The transverse forces in a leg segment are usually measured by strain gauges which detect bending. In the case of the OSU hexapod, a strain gauge is bonded to each of two adjacent faces of the square-sectioned leg. A semiconductor rather than metal strain gauge is used because of its greater sensitivity, since the forces and resulting deflections are low.

The measurement of axial force requires some kind of load cell to be built into the leg. These may use strain gauges bonded to a metal ring which deflects under load, or a piezoelectric sensor. If the load cell is sensitive to bending moments or transverse forces it must be isolated from these by, for example, fitting it inside a telescopic joint (Klein *et al.* 1983). Multiaxis force sensing has also been investigated for industrial robots (eg Gaillet and Reboulet 1983).

TOUCH SENSING

The simplest touch sensor is a microswitch with a mechanical actuation arrangement. Such sensors have been used to detect foot-ground contact. Home-made mechanical switches with special geometries have also been used to, for example, detect collision of a foot with a step or obstacle (Hirose 1984). If sensitive to a sufficiently small load, the leg force sensors may also be used for such a function.

Touch sensing has been investigated more thoroughly in the context of manipulators. Piezoelectric polymers seem to have a potential as a continuous touch sensitive surface (Bardelli *et al.* 1983, Benjamin 1983, Dario *et al.* 1983). Numerous other techniques, both for simply switching on contact and for pressure measurement, have been investigated, eg inductive, capacitance, resistance and optical transducers.

PROXIMITY SENSING AND SHORT DISTANCE RANGE MEASUREMENT

Proximity sensing can be defined as a binary indication of the presence of an object within a certain sensitive volume. Its uses are in collision avoidance, manoeuvring in confined spaces, controlling foot placement, and guiding a foot up a step or over an obstacle.

The most versatile medium for distance measurement in the range of up to 1m is ultrasonics. Pulsed sonar at the commonly used frequency of 40kHz has a distance resolution of about 1cm, and a typical pulse repetition rate is 10 pulses/s (Witkowski *et al.* 1983). The sensors are small and cheap and so can be used in numbers. Sonar can be used for both ranging and proximity sensing.

Proximity sensing can also be done with infrared sensors (their range is a few centimetres). They have been used, for example, to surround the arm of a manipulator with a sensitive 'cuff' (Witowski *et al.* 1983). The sensitive volume is defined by the intersection of a cone of radiation emitted by an infrared LED with the field of view of a detector diode. A group of emitters and receivers may be used to shape the sensitive volume.

BODY ORIENTATION AND ANGULAR RATES

The terms 'attitude' and 'orientation' are often interchangeable, but it is sometimes helpful to distinguish yaw or azimuthal angle from pitch and roll. On a level surface yaw is distinguished because it is a navigational quantity, rather than reflecting an internal disposition. Here the term attitude is used to mean pitch and roll, and azimuth or compass orientation is used for angle about the vertical or surface-normal axis. Orientation by itself is taken to include all three axes.

There are various ways of expressing orientation, for example Euler angles. Choosing a consistent system of reference is important if the changes in angle are large, but it is often adequate to use terms such as pitch and roll rather loosely. As used here pitch is defined as the angle of rotation about a transverse axis fixed in the body, and roll is the angle about a longitudinal axis fixed in the body.

The usual sensors of pitch and roll are gyroscopes or damped pendulums. Pendulums can be relatively cheap but are sensitive to acceleration. They have been used on the OSU hexapod and the Tokyo Institute of Technology quadruped PVII. They can be instrumented to give a continuous measure of angle, or to give just a binary indication of departure from the vertical by a certain amount. One solution to the problem of their sensitivity to acceleration is to mount one on each foot and use the signal only from those known to be firmly on the ground. This, of course, assumes that the relative orientation of body and foot is known.

A better instrument (a typical accuracy is $\pm 1°$) is the vertical gyro, which provides both pitch and roll signals from a single instrument, usually as potentiometer outputs.

Angular rate in pitch and roll may in principle be obtained by differentiation if the angle is known with enough accuracy, but if the rates are high (tens of degrees each second) or if high accuracy is needed it is better to use a rate gyro. A separate gyro is needed for each axis.

For azimuth, gyroscopic sensors are available but expensive. An alternative is the fluxgate magnetometer. These are commercially available or can be built without too much difficulty (Pollock 1982). An accuracy of about 1° is possible at mid latitudes away from large masses of iron. Another method is to sense orientation relative to a beacon or set of beacons, which can also give other information such as position. The beacons may be optical, acoustic or radio depending on the range and propagation conditions. A beacon system is obviously only useful when it is possible to place beacons in advance — it has been used for the underwater navigation of submersibles, for instance.

BODY POSITION, SPEED AND ACCELERATION

These quantities have a global and a local interpretation. On a global scale body position means map position and its determination is navigation, an issue not specific to walking machines. Speed, for navigational purposes, is speed averaged over some fairly long period, and acceleration is not directly relevant.

On a local scale these quantities are aspects of the problem of achieving smooth and stable locomotion. The most direct way of sensing body position is by finding the range to known objects. A second way is to calculate the coordinates of a body reference point from the measured joint angles and the positions of the supporting feet. It is likely that the foot positions will not be known in an absolute sense, but at least this enables the position of the body relative to the local surface to be determined.

Velocity and acceleration can again be calculated from the joint velocities and accelerations, but it is often preferable to sense them directly. Velocity can be sensed by Döppler radar or sonar aimed at the ground ahead; or, in principle from analysis of television pictures, although this is not yet an established technique. Acceleration in the range relevant to dynamic control can be sensed by accelerometers.

TERRAIN SCANNING

A mobile robot, whether legged or not, needs to scan the ground ahead to avoid obstacles and, if it has legs, select footholds. Terrain scanning can perhaps be dispensed with on smooth ground, but this is probably a rare condition for legged robots. It is also unnecessary if it has a human driver who is prepared to control foot placement but, as with riding an animal, it is better if the driver can concentrate on other things while the machine chooses its own footholds. If a walking vehicle is an autonomous robot then terrain scanning becomes connected with navigation, as the machine must select not just a locally suitable path but an effective route to its destination.

Because of the range and resolution requirements, terrain scanning is usually done at visible or near-infrared frequencies rather than by sonar or radar. There are several possible methods, including:

1) monocular scene analysis (using a single television camera),
2) binocular scene analysis,
3) triangulation ranging using a television camera or other detector to locate the spot of light where a laser beam meets the ground,
4) time of flight ranging,
5) interferometric (phase comparison) ranging.

These methods all have limitations. In particular, the scene analysis methods, although the subject of much research, are far from being a mature sensing method for reliable use on a walking machine.

Perhaps the most sophisticated system yet to be designed for a walking machine is the continuous wave phase comparison system for the OSU Adaptive Suspension Vehicle (Zuk 1983; Waldron *et al.* 1984). This principle of range measurement is used in surveying equipment. A beam, in this case of infrared light from a gallium arsenide laser, is amplitude modulated and phase comparison of the transmitted and reflected beam gives the range. The maximum range is about 10m (although by using multiple modulation frequencies the technique can be extended to a range of kilometres) and the resolution a few centimetres. The beam is scanned in a raster pattern covering the ground ahead of the vehicle.

SOIL PROPERTIES

People and animals can usually judge quite accurately when the ground is slippery or otherwise dangerous. It is possible to predict, when one's feet sink into sand, snow or soil, whether the depth of sinkage is limited by compaction or not; and to evaluate qualitatively whether walking on a particular surface dissipates a lot of power.

These abilities will be needed in an effective rough-country vehicle, but so far little attention has been paid to their achievement. As mentioned earlier it is possible from leg forces to detect slipping, and some other ground properties can be inferred from the standard instrumentation; for example, information on how the soil yields and how much power is dissipated can be obtained by recording the history of leg force and movement after the moment of ground contact. To distinguish foot sinkage from other movements in the system it may be necessary to fit extra instruments, such as an accelerometer on the foot. Another way would be to combine joint angle measurements with sonar body height sensing.

It might also be worth designing instruments specifically for soil properties; for example, a collar round each leg, which would be pushed up the leg by the soil as the leg sank in, could be fitted with a displacement transducer to measure foot sinkage. There are many other possibilities, such as probes to measure ground hardness, temperature and moisture, which would give indirect indications of the likelihood of slipping or sinking.

The Design of Algorithms for Multipod Walking

This section discusses the design of algorithms for multipod walking and presents examples which show how a hierarchical arrangement of the processor architecture or software structure tends to arise, and which therefore implies requirements for programming languages and software environments (ie an operating system, monitor or executive allowing real-time running of a robot control program). It may also be regarded as a more detailed exposition of the section on gait generation (see page 121).

Algorithms or equivalent descriptions have been presented for several gaits. Examples are free gaits — McGhee and Iswandhi 1979, Hirose 1984; wave gaits with turning — Orin 1982; follow-the-leader gait — Ozguner *et al.* 1984.

The first example presented here is a simplified model of a wave gait implementation for straight line walking on level ground. As mentioned in Chapter 4, such a task can be split into several levels — as shown in Figure 5.9. The bottom level, joint servo control, is too detailed to discuss here. Joint torque planning is a refinement which is often omitted; it is concerned with functions such as equalizing the load on the legs. The top two levels are more fundamental. The diagram refers to a multipod with legs of the geometry shown in Figure 5.10; for simplicity the third joint needed for vertical foot movement has been ignored. Planning the return phase of foot movement can also be ignored; its path is of only secondary interest (for a detailed example of foot motion planning see Orin 1982). It can be seen from this model that the levels are essentially separate processes all running simultaneously and communicating various kinds of data, suggesting the architectural model of Figure 5.1.

Mention has been made previously of finite state control (see, in particular, Figure 1.9) and it is worth seeing how this fits into the hierarchical model. In its simplest form, the finite state controller combines gait generation and output to the actuators in a single level, eliminating separate levels for motion planning, torque planning and joint servo control. This has the merit of simplicity of architecture and program structure, although at the cost of restricted performance. An example of a finite state controller which produces the alternating tripod gait for a hexapod is shown in Figure 5.11 (Todd 1984). The two

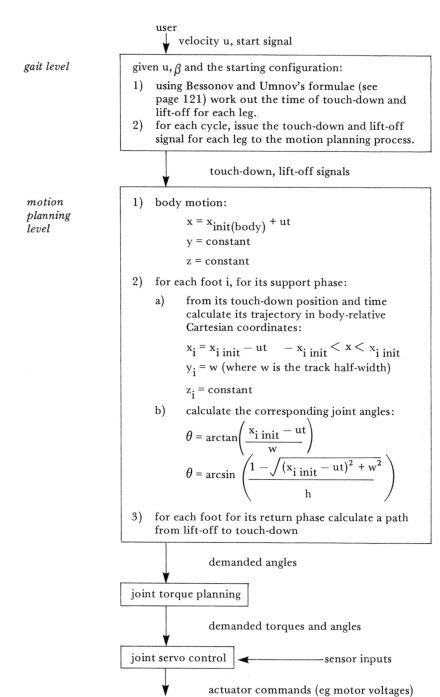

user
velocity u, start signal

gait level

given u, β and the starting configuration:

1) using Bessonov and Umnov's formulae (see page 121) work out the time of touch-down and lift-off for each leg.

2) for each cycle, issue the touch-down and lift-off signal for each leg to the motion planning process.

touch-down, lift-off signals

motion planning level

1) body motion:

$$x = x_{init(body)} + ut$$

$$y = \text{constant}$$

$$z = \text{constant}$$

2) for each foot i, for its support phase:

a) from its touch-down position and time calculate its trajectory in body-relative Cartesian coordinates:

$$x_i = x_{i\ init} - ut \qquad -x_{i\ init} < x < x_{i\ init}$$

$$y_i = w \text{ (where w is the track half-width)}$$

$$z_i = \text{constant}$$

b) calculate the corresponding joint angles:

$$\theta = \arctan\left(\frac{x_{i\ init} - ut}{w}\right)$$

$$\theta = \arcsin\left(\frac{1 - \sqrt{(x_{i\ init} - ut)^2 + w^2}}{h}\right)$$

3) for each foot for its return phase calculate a path from lift-off to touch-down

demanded angles

joint torque planning

demanded torques and angles

joint servo control ← sensor inputs

actuator commands (eg motor voltages)

Figure 5.9 *Hierarchical structure of hexapod wave gait algorithm.*

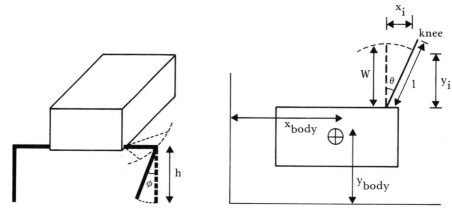

l = length of upper leg segment

$x_i\ y_i$ = body-relative coordinates of foot (not knee)

Figure 5.10 *Definition of terms used in foot movement calculations.*

sets of legs which alternately support the machine are here labelled 'red' and 'blue'. The machine can reverse at any time. The cycle shown here is a very slow one, as the machine pauses while one set is lowered and then the other is raised. The flow chart of the program which implements this finite state machine is shown in Figure 5.12.

The more complex control method of model reference control supervised by finite state control (see Chapter 4) fits more clearly into the hierarchical pattern: the finite state controller is the gait generator, then there is a level of motion planning and finally, a level of joint servo control.

Programming Languages for Legged Robots

Up to now nearly all legged robot control programs have been written in general purpose languages such as Pascal and Fortran, and in assembly language. Only one language has been designed specifically for walking machines, and this is not in wide use. Yet in the related field of industrial (manipulator) robots there are dozens of languages in use. So it seems worth asking why this situation has arisen and what conclusions are to be drawn about the future of programming for legged robots.

Manipulation robot languages have been reviewed by Lozano-Perez (1982), who discusses the requirements for and the classification of

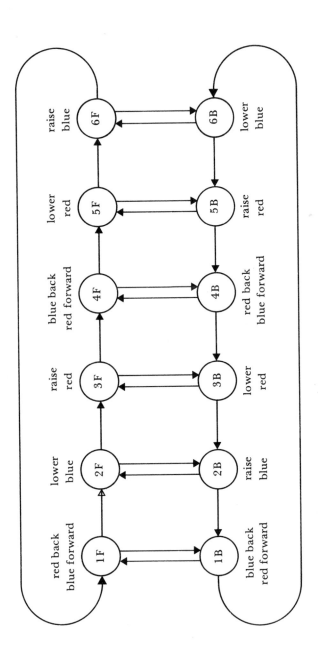

Figure 5.11 *Finite state machine implemented by the control program of a hexapod. Each state represents an action and the arrows show which states can be reached from it. The two sets of three legs used in the alternating tripod gait are labelled 'red' and 'blue'.*

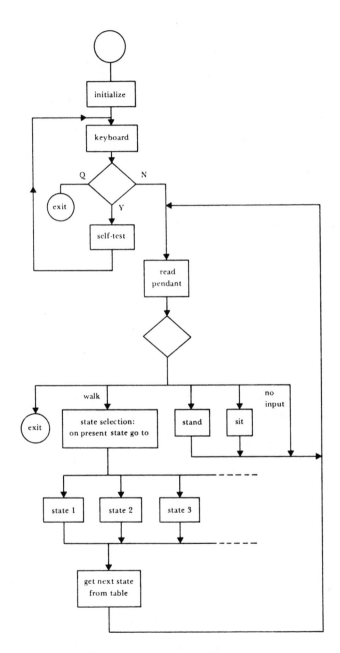

Figure 5.12 *Flow chart of a hexapod walking program.*
After initialization and a self-test stage, the main loop is entered
in which the operator's controls are interrogated
and one state of the cycle is executed.

robot languages. The requirements include sensing, world modelling, motion specification, flow of control and programming support.

Sensors are used for starting and ending motions, for choosing a particular action among alternative actions, for identifying and locating objects and their features, and for complying with external constraints. A language should provide suitable input/output features and also cope with the indirect consequences of altering actions in accordance with variable sensed data.

World modelling is concerned with representing the positions of objects and their features, including the components of the robot itself. It is frequently necessary to transform between several coordinate systems. The programming system must provide adequate facilities for the input, representation and computation of geometrical information.

Motion specification is the primary object of robot programming — it is telling the manipulator to go to a point or follow a path. In some cases it is enough to specify a few points; in others the shape of and even the speed along a continuous path must be given. A further issue is whether or not some joint movements can be left unspecified when there are alternative equivalent configurations. A programming system will have to provide for some of these ways of specifying motions, and possibly others, such as allowing the user to input a mathematical function of time.

A language must take into account flow of control. In addition to the need for loops and branches on sensor tests, there are more complex issues such as the coordination of concurrent processes.

Finally, the system should have adequate facilities for rapid program modification and debugging, and perhaps simulation, and should have interfaces to other systems.

Lozano-Perez goes on to classify robot programming systems into three categories: guiding (leading the arm through the required motions and recording them), robot-level systems in which the user writes a program in terms of robot movements and sensor tests, and task-level systems in which operations are specified by their desired effect on objects, leaving the system to translate these into robot-level actions. He surveys more than 20 languages, most of which are in the robot-level class.

The earliest robot-level language was MHI, developed in 1960 at the Massachusetts Institute of Technology. Its main robot-specific constructs are moves and sensor tests, such as *move* (parameter) *until* (sensor condition). A more general purpose language was WAVE, developed at Stanford in the early 1970s. It introduced the description of positions by Cartesian end-effector coordinates, coordinated joint motions, and compliance by letting certain joints move freely under external loads. (If the joint is not back-drivable this requires servo control with the external force as an input to be minimized.)

Perhaps the most influential language, and one which is still being extended, is AL. An AL program is compiled on one computer into a lower level language which runs on the second, real-time control, machine. AL provides Cartesian specification of motions, and compliance, the data types and control structures of an Algol-like language, support for world modelling (for example, there is an affixment function), and the concurrent execution of processes, allowing synchronization.

Task-level languages are more a subject of research than an available tool. It is necessary to be able to model the world of robot and objects in the language; to specify a series of models, possibly just two, to represent the stages which the assembly or other task must go through; and for the system to plan a series of movements. The subject merges with planning and problem solving as a branch of artificial intelligence.

It is now worth asking in what ways the programming requirements of locomotion differ from and resemble those of manipulation. One difference is that an industrial robot is an automaton which once started requires no human intervention, whereas a legged robot is a vehicle with (usually) a driver. A walking machine must interpret and execute commands which may contain variable parameters such as rate of turn; also, its actions may vary according to the terrain. This implies that guiding in the industrial robot sense is of little use in locomotion. Another difference is that concurrency is less of a problem with an industrial robot since in most cases actions are sequential (although requiring the simultaneous motion of several joints). A multi-legged robot, on the other hand, often needs several processes which overlap unpredictably — the situation is analogous to a multi-manipulator installation rather than to a single arm.

Turning to similarities, an obvious requirement common to manipulators and legged robots is spatial calculations, including coordinate transformations. There are, of course, more basic necessities such as sensor conditionals. In some cases compliance in the sense of allowing free joints may be desirable. Another similarity is between planning a route, or a series of foot placements, and planning a manipulator task. Indeed, certain functions, such as planning a route for an object among obstacles, may be both very similar in the two cases and difficult to accomplish. Taking these comparisons into account it appears that languages at least as powerful as AL are desirable for functions such as leg coordination and foot motion planning.

Donner (1983) concluded that AL's mechanisms for handling concurrency could be improved on, and went on to design a language called OWL specifically for walking machines, in particular Sutherland's hexapod (see page 157). One of its goals is to minimize latency, that is, the delay in responding to an event. It does this by using very small blocks of code each guaranteed to take less than some fixed amount of time. The OWL compiler is written in C using YACC, a technique which allows the relatively easy development of new languages. It generates C which is then compiled and run on a VAX or the hexapod's on-board 68000.

Most legged robot control programs continue to be written in conventional languages; an example was given in the section on architectures. This probably reflects their greater availability and familiarity rather than a belief that they are better than languages such as AL. This probably also explains why real-time languages such as Modula and RTL/2 have not been used. However, as legged robots become more common it seems likely that more attempts will be made to introduce languages providing the features discussed here.

Chapter 6
Current developments

The main subjects involved in the design of walking robots have been discussed in previous chapters. This chapter describes some of the projects active today and may be regarded as an update of the details discussed in Chapter 1.

The Ohio State University (OSU) Hexapod and Adaptive Suspension Vehicle (ASV)

Research with the hexapod is continuing. As far as the drive towards practical vehicles is concerned the ASV will probably be the leading machine of the next few years. The project draws on many years' experience with OSU's electric hexapod and is evidently on a scale which allows not only the extensive use of components such as variable displacement pumps and servo valves, but also the development of special devices such as light-weight actuators, a terrain-scanning system and an energy storage flywheel.

The ASV is described (Waldron *et al.* 1984) as a 'proof of concept prototype' of a proposed class of rough-terrain vehicles. It is a hexapod and uses statically stable gaits. It is 5m long, 3.3m high, 1.6m wide and weighs about 2600kg (see Figure 6.1). It carries a driver and is designed for a payload of 225kg; the maximum intended speed is 3.6m/s. The machine is powered by a 0.9-litre, 70kW motorcycle engine. This drives a 0.25kWh (0.9MJ) flywheel and then a set of 18 variable displacement hydraulic pumps by means of shafts and toothed belts.

The coordination of foot placement and leg movement and the control of body attitude is handled by a network of 13 Intel 86/30 microcomputers and is usually automatic, freeing the driver to attend to the more strategic decisions about overall speed and direction. The operator can have continuous simultaneous control of forward speed, lateral speed and rate of turn, and can also set required values for body height and attitude.

The machine uses vertical gyroscopes, rate gyroscopes and accelerometers to sense body movement and attitude; all legs are fitted with proximity and three-axis force sensors. An interferometric range finder

Figure 6.1 *The Ohio State University Adaptive Suspension Vehicle.*

(see Chapter 5) scans the ground ahead of the robot.

The focal point of the project is the attempt to push speed and efficiency much further towards those needed for a practical vehicle. Machines like the OSU electric hexapod are an order of magnitude too slow and inefficient to be a suitable model. Therefore the ASV required some radical changes of design.

From Figure 6.1 it can be seen that the legs swing in the sagittal plane like those of a mammal. This arrangement has the advantage of allowing a relatively narrow vehicle, and permits low twisting moments about the leg sections. Further, the near-vertical alignment of the leg reduces bending moments in the upper leg segment. It also allows a relatively long stride if the movements of knee and thigh are coordinated properly. However, the OSU team's main criterion was energetic efficiency and they felt that the advantage of this geometry was the energy saving due to the pendulum-like character of such a leg. On each stroke gravitational energy can be converted to kinetic energy and back again.

However, as discussed in Chapter 2 (see Figure 2.5), this configuration in its most simple form suffers from the inefficiency known as geometric work. To overcome this problem a pantograph (see Figure 6.2)

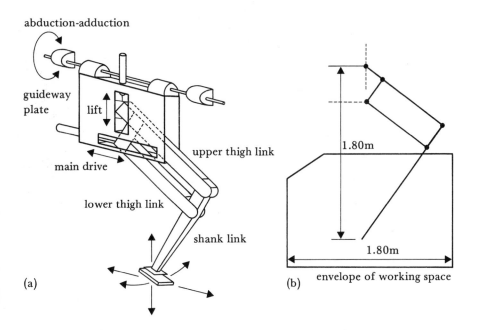

abduction-adduction

guideway
plate

lift

main drive

upper thigh link

lower thigh link

shank link

(a)

1.80m

1.80m

(b) envelope of working space

Figure 6.2 *The pantograph leg mechanism of the ASV:*
(a) shows how the pantograph assembly pivots about a longitudinal
'abduction/adduction' axis; (b) shows an example of a typical working envelope
for such a pantograph; it is the intersection of a rectangular desired working space
with the reachable space (the corner is cut off
because of the limits on the joint angles).

is used to separate the vertical and horizontal motions and transfer them from the actuators to the foot. The pantograph also allows a long stride; the length of the leg's walking envelope is about equal to the leg length measured from hip to ground (see Figure 6.3). In the absence of a pantograph such a long stride would require coordinated motion of knee and thigh over a range of angles for which the force level and direction would vary considerably, giving a difficult control problem.

The disadvantage of this approach is that it requires sliding joints, which are always difficult to design satisfactorily. In this case, a design was evolved which uses pairs of conical rollers running on V-profile tracks. This is shown in a simplified form in Figure 6.3. Each sliding joint uses a pair of guideways, one on each side as shown in Figure 6.3b. This eliminates any bending moment on a guideway. The thigh link pivots on a block which runs on eight rollers, four on each side to engage with a guideway. The rollers of each vertical pair are in contact with each other as well as with the guideway. This allows the load on the joint to be shared among all the roller shafts. The design is described in detail elsewhere (see Vohnout *et al.* 1983).

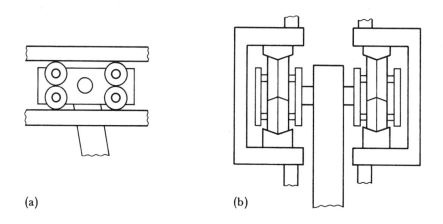

(a)   (b)

Figure 6.3 *The guideways of the ASV's sliding joints:*
(a) side view showing how a link of the leg is pivoted on a carriage
with rollers on four shafts; (b) cross-section showing V-profile of the tracks
and the conical rollers which run on them.

Although sliding joints are used for the propulsion and lift motions, a rotary joint with a horizontal fore-and-aft axis is used for the third degree of freedom, responsible for lateral movement of the foot. Waldron refers to this motion as abduction/adduction. As can be seen from Figure 6.2, the double guideway assembly or slider box pivots about the abduction/adduction axis.

The second major area of innovation is power transmission and actuation. For the reasons discussed in Chapter 3, at present hydraulic transmission is the best choice for a large free-ranging vehicle. The most usual kind of hydraulic transmission uses servo valves to control the flow of oil from a central pump to the actuators. This approach was extensively studied (Gardner *et al.* 1983) but it has the disadvantage that both the lift and propulsion actuators are often required to move quickly but under only light loading, a condition leading to excessive power dissipation in the valves. This happens because the pressure must be kept up for other actuators in the system which are supporting heavy loads, so there is a large flow at a high pressure drop through the valves controlling the lightly loaded actuators.

Ways of avoiding this problem are possible, but they involve extra

pumps and valves with their associated weight and cost penalty. Therefore, hydrostatic systems were re-evaluated and it was concluded that
this approach would give better performance. The type of circuit used
for each joint is illustrated, in a simplified form, in Figure 6.4.

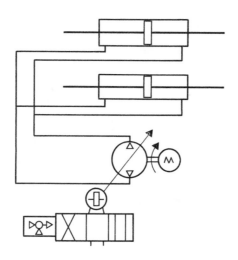

Figure 6.4 *Hydrostatic hydraulic circuit for one joint of the ASV.*

In the case of the lift and propulsion joints a pair of actuators is fitted,
one on each side of the joint, to even out the load. Each actuator is an
equal area cylinder, a necessity for a hydrostatic circuit. The piston rod
is fixed to the frame (to the slider box in the case of the lift and propulsion actuators) and the cylinder barrel to the sliding assembly. Oil is
supplied through the piston rod (see Figure 6.5). The swashplate angle
of the variable displacement pump is set by a small rotary hydraulic
actuator. This is controlled by a servo valve and powered by an oil
supply generated by a separate pump common to the whole system.

The three variable displacement pumps for each leg are driven by
toothed belts from a shaft, common to all three legs on one side,
which runs through the centre of the abduction/adduction bearings.
The propulsion pump and lift pump are mounted on the slider box and
the abduction/adduction pump on the body. The two shafts are driven
by the engine which is also coupled to the flywheel. The flywheel is
important because the pumps present strongly varying loads; it also
allows regenerative braking with the pumps acting as motors backdriven by the actuators, at least in theory.

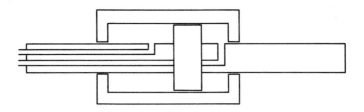

Figure 6.5 *An ASV linear hydraulic actuator:*
the rod and piston are fixed; the cylinder slides along them.

ReCUS: The Komatsu Underwater Octopod

This machine is of interest because it shows that there is a future for machines which, while advanced in some respects, avoid almost completely the issues which were seen as important by the designers of the ASV. The reason this is possible is, of course, that it is intended for a completely different function and environment: its principal design criteria are stability and accuracy of positioning. Speed and energetic efficiency are of only secondary concern.

The 'ReCUS' robot has been developed by Komatsu Ltd for surveying the sea-bed prior to bridge pier and breakwater construction by the Honshu-Shikoku Bridge Authority (Ishino *et al.* 1983). It is shown in a simplified form in Figure 6.6. The inner frame, with four vertically telescoping legs, slides longitudinally on the bed of the outer frame which also has four legs. To walk, the machine stands alternately on the outer and inner leg sets: an alternating tetrapod gait. The steering method is described in Chapter 4 (see Figure 4.17).

ReCUS (Remotely Controlled Underwater Surveyor) is 8m long, 5m wide, 6m high and weighs 29 tons in air. The step length is 2.5m, the leg cylinder stroke is 2.2m and the walking speed is in the range 1.3 − 4m/min. It has been used in depths of 70m. The mother ship is a floating crane which lowers the robot to the sea floor. It carries a control cabin and the generator to supply power to the 37kW three-phase motor which drives the hydraulic pumps.

Walking is controlled by a PDP-11/03 single board computer on the mother ship, communicating with the ReCUS by a cable. A vertical gyro is used to sense body attitude. On a rough sea-bed a specified attitude can be maintained within ±1°. A typical sea-bed roughness is 1 − 2m. The horizontal attitude can be maintained on slopes of up to

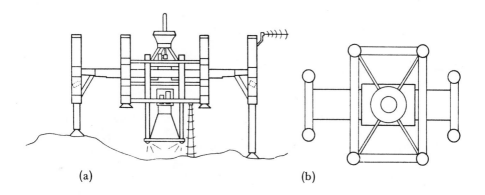

(a)   (b)

Figure 6.6 *The underwater octopod robot ReCUS:*
(a) side view — the cone at the top is a docking device for the lifting gear;
the cone at the bottom is a clear viewing aid; (b) plan view showing
the central pivot which allows the inner frame to turn relative to the outer one.

20°. Position in the horizontal plane is determined using ultrasonic transponders fixed to the sea bed.

For its surveying function the robot carries a collection of television and still cameras and an ultrasonic scanning system. Komatsu have also proposed that ReCUS could be used for construction jobs such as levelling rubble. It would be fitted with a rake sliding on the main frame. Other proposed applications are drilling and trench digging.

A smaller but geometrically similar machine with magnetic feet for crawling over the vertical surface of a steel pressure vessel has been described by Kemmochi and Kazuoka (1983). Its use is ultrasonic inspection.

Sutherland's Hexapod

This machine, designed by I.E. Sutherland of Carnegie-Mellon University and Sutherland, Sproull and Associates, is significant in being the first man-carrying computer-controlled walking machine (Raibert and Sutherland 1983; Sutherland and Ullner 1984). Its design is also interesting for its use of a leg geometry and hydraulic circuit design intended to reduce the control burden on both computer and driver by automatically coordinating joint motions in ways suitable for walking.

The hexapod, whose basic geometry is shown in Figure 6.7, is about 2.5m long and the same width. It weighs about 800kg and is powered by a 13kW gasoline engine driving four variable displacement pumps. The walking speed in the alternating tripod gait is 0.1m/s. It can also walk sideways at rather more than half this speed.

Figure 6.7 *Sutherland's hexapod.*

It has an unconventional arrangement of its hip actuators. Two cylinders are mounted in a 'V' above the leg. It is possible to set the valves so that as one shortens the other lengthens in such a way as to produce horizontal movement, whereas if they both move in the same sense they move the leg vertically. (A third actuator for the knee produces sideways movement.) This hip arrangement is one instance of what Sutherland calls a 'passive hydraulic circuit'. Such circuits achieve joint coordination in two ways. First, actuators are sometimes connected together in series so that as oil flows out of one it must flow into the next. This forces the actuators to move the same amount. Second, if two or three are connected in parallel they will automatically share any applied load equally. In this connection their collective movement is actively controlled by a pump, but their differential movement is determined only by the relative loading of the actuators.

The hip connection for horizontal movement consists of putting two actuators in series so that as oil flows out of the fixed end of one it flows into the fixed end of the other. The motion is exactly horizontal only if the plane of the cylinders is inclined at 45° (Sutherland and Ullner 1984). A series connection is also made between the front and back actuator-pairs on each side to coordinate their movement during the propulsion stroke. This arrangement is shown for one tripod-set of legs in Figure 6.8. In this illustration legs 2, 3 and 6 are being driven together. Each side has a separate pump so only legs 2 and 6 are connected together; the coordination between this pair and leg 3 is achieved by non-hydraulic means. Other series connections are possible.

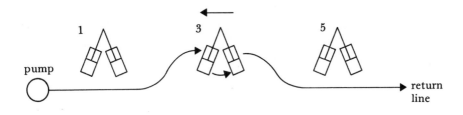

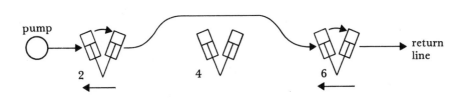

Figure 6.8 *Series connection of a hydraulic circuit for the propulsion stroke of Sutherland's hexapod.*

The parallel connection is used for various purposes such as raising and lowering a set of legs, and to connect the knee actuators. If the three knee cylinders of a supporting tripod-set of legs are connected in parallel then although their collective sideways movement can be controlled by a pump, their differential movement is free and compensates both for the movement of the knee in an arc during forward rectilinear walking and for the larger sideways knee movement which must occur during a turn. This knee coordination is perhaps the most successful application of a passive hydraulic circuit.

The valves are all directional, not proportional or servo, spool valves, which are relatively cheap and simple but cannot control speed. Speed is regulated by manual control of the displacement of the pumps. The combination of the main propulsion pump and sideways motion pump flow rates governs the speed and direction of walking and the rate of turn. The pump displacements are controlled by pedals and a joystick.

The role of the on-board computer is to switch the valves on and off in the sequence appropriate to the specified gait. It can interrogate joint angle and leg force sensors, and the driver's controls. Several types of program have been written to test different methods of control, and a special language (OWL) has been developed (Donner 1983). The robot, which was built as a way of learning about hydraulic actuation, has now been scrapped.

The Odetics ODEX I

This machine is recognizable by its startling appearance, with a spherical transparent dome surmounting a vertical cylindrical body (Russell 1983). The six legs are disposed symmetrically about the vertical axis, an arrangement previously used only by Ignatiev (Vukobratovic 1973). The general appearance of the machine is shown in Figure 6.9a and the leg design in Figure 6.9b. The vertical and horizontal, or extension, drives use leadscrews and the third joint, the vertical axis of rotation or 'swing' uses a spur gear train. In all cases the drives are electric. The power source is an aircraft battery.

As mentioned in Chapter 3, the robot can adopt a range of postures. Its height varies between about 1.2 and 3m. It weighs 200kg. One of its main features is its great strength: a single leg used as a manipulator can lift about 200kg and the whole machine can walk while carrying a load of around 400kg.

The significance of ODEX I is mainly that it shows what can be done with an unconventional geometry. Its symmetry and variable aspect make it manoeuvrable in confined spaces. The use of legs as manipulators (under teleoperator control, not automatically) is also of interest. This leg arrangement does not appear suited to high speeds.

Quadruped Research at the Tokyo Institute of Technology

Over the last ten years or so Hirose and Umetani and their co-workers have devised a variety of ingenious robotic mechanisms, such as the 'active cord' described in Chapter 1. One branch of this work has been concerned with walking robots which are agile on rough ground and also energetically efficient. They have built a series of electrically powered quadrupeds which use pantographs and related mechanisms for efficiency. One of these is shown in Figure 3.15a and is more fully described in Hirose and Umetani (1980).

Among the latest machines in this series are the PV II (Hirose 1984) and the TITAN III (Hirose *et al.* 1984). PV II is shown in Figure 6.10;

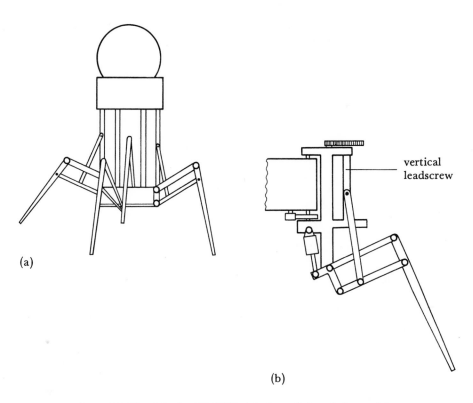

vertical
leadscrew

(b)

Figure 6.9 *The Odetics ODEX I: (a) view of the whole machine;*
(b) the linkage of one leg showing how the vertical and extension motions
are driven by motors with leadscrews. The third motion, rotation about a vertical
axis, is driven by a motor and a spur gear train.

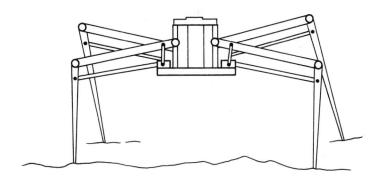

Figure 6.10 *The leg arrangement of the PV II.*

it is about 1m high and weighs 10kg. The leg design incorporates a three dimensional pantograph (see Figure 6.11). It uses three linear actuators (d.c. servomotors with leadscrews). Hirose estimates that at low speeds on level ground the decoupling of the lift and horizontal movements reduces energy losses by up to 80 per cent. The wire and pulley linkage is added to keep the foot vertical.

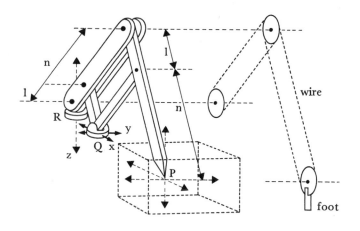

Figure 6.11 *Pantograph used by PV II.*

The PV II has been used to test several control ideas, some of which are quite simple, such as keeping the body level by adjusting the leg lengths in response to signals from an on-off pendulum sensor. An obstacle contact reflex can lift the leg over a step (see Figure 6.12). It has also been used, at least in simulation, to study the problems of walking on surfaces with forbidden zones. Hirose has devised what appears to be a flexible strategy for selecting footholds while maintaining, as far as possible, a regular gait. TITAN III is similar but more sophisticated; it uses carbon fibre legs (see Figure 6.13).

Walking Robots in France

A group at the University of Paris has built a simple electric hexapod (Kessis *et al.* 1983). It uses the linkage shown in Figure 6.14 to decouple the vertical and horizontal motions and to allow the motors to be mounted on the body. The problem of turning is dealt with not by a third actuator in each leg but by making the legs flexible laterally so they can bend during a turn. This approach seems likely to be satisfactory only on a level surface.

Figure 6.12 *PV II walking on an uneven surface.*

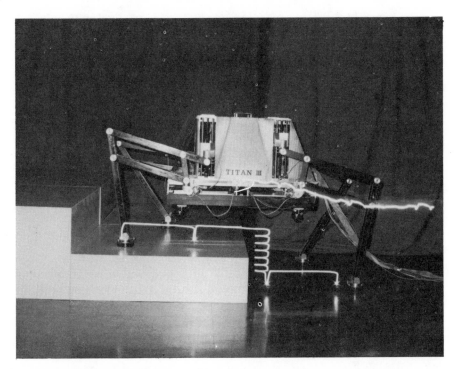

Figure 6.13 *TITAN III climbing a step using an obstacle sensing reflex.*

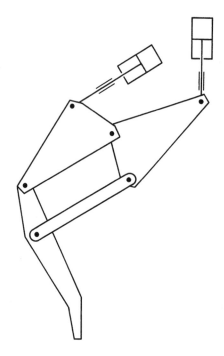

Figure 6.14 *A linkage for decoupling vertical and horizontal motions.*

Jurassic Robots

The aims of Jurassic Robots are to develop simple walking machines of limited performance which can be made relatively cheaply, avoiding expensive components such as servo valves; and to explore applications of these. Pneumatic and hydraulic hexapods have been built; a pneumatic machine is shown in Figure 6.15. Current work includes research into the use of balancing devices in walking on slopes; and a design study of sauropod reconstructions. This type of application is discussed in Chapter 7.

Current Research on Bipeds

Some of the work being conducted at Waseda University is summarized in Chapter 1. Recently, other research on bipeds has become active in Japan. Since about 1980 a group at Osaka University has been working on hierarchical control structures which allow dynamically stable walking, and has built bipeds to test this method (Miyazaki and Arimoto 1983a, b). Another series of bipeds for dynamic walking experiments has been built at the University of Tokyo. One of these, called BIPER-4 (see Figure 6.16), has seven powered joints: a hip roll,

Figure 6.15 *A pneumatic hexapod (without on-board compressor)
in an early stage of development.*

ankle pitch and knee joint in each leg and a hip pitch joint connecting
the two legs (Miura and Shimoyana 1984). The actuators are d.c.
motors. The instrumentation consists of conductive plastic foot contact
sensors and joint angle potentiometers. Dynamically stable walking of
BIPER-3 has been achieved. This biped has no knee joints and its ankle
joints are not powered. The feet do not contribute to pitch stability
but allow the angle of the leg from the ground plane to be measured.

Research on Hopping at Carnegie-Mellon University

The importance of balance to legged locomotion has long been evident.
The earliest experimental work on balance was Cannon's control of
inverted pendulums balanced on a powered truck (Higdon and Cannon
1963). The most extreme case of balance in locomotion is leaping on a

single leg, which, if the method is to be at all practicable, must be springy like a pogo-stick. This ballistic method of locomotion has been proposed for lunar travel (Seifert 1967), and Matsuoka (1980) has built planar one-legged hopping machines.

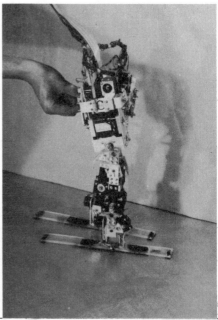

Figure 6.16 *BIPER-4 showing hip roll, ankle pitch and knee joints; the legs are connected by a hip pitch joint (photograph by Professor H. Miura).*

However, by far the most advanced research is that of Raibert and his colleagues at CMU's Robotics Institute. Their experimental work began with a one-legged hopper confined to a vertical plane (or rather to a spherical surface of large radius, since the confinement was by tethering it by a beam to a central pivot). It was in principle similar to that shown in Figure 6.17 except that the leg swung in a plane instead of being gimballed, and the hip actuator was pneumatic. It succeeded in hopping stably and leaping over an obstacle (Raibert *et al.* 1983a). The next major experimental phase was the development of the three-dimensional hopper (its control method was described in Chapter 4). A photograph of it is shown in Figure 6.17. As with the two-dimensional machine its weight is kept to a manageable level by supplying compressed air, hydraulic oil and control signals through an umbilical cable from equipment in the laboratory.

The hopper runs at up to 2.2m/s with strides of up to 0.79m (Raibert *et al.* 1984). When hopping on the spot it keeps its position to within

Figure 6.17 *The Carnegie-Mellon three-dimensional hopper in action [reproduced by the kind permission of M.H. Raibert from* International Journal of Robotics Research — Special Issue on Legged Locomotion *1984, 3 (2)].*

0.25m. It has also succeeded in hopping round a square path with reasonably sharp turns.

Following the success of this machine the research group have turned their attention to four-legged running machines. Computer simulation of a planar two-legged model (thought of as the lateral half of a quadruped, and referred to as the 'planar dog') has shown (Murthy in Raibert *et al.* 1983b) that it is possible to control a trot-like gait in which the two legs thrust in phase, with the differential thrust being used to regulate body pitch. It has also revealed that if the relative phase of the leg thrusts is not predetermined then the machine settles into a stable pitch oscillation (regardless of whether it is travelling or bouncing on the spot). This raises interesting questions about what gaits are self-stabilizing in multilegged animals (and machines). Experimental work also is now under way. A four-legged running machine is being developed (Brown and Raibert in Raibert *et al.* 1983b) using telescoping hydraulic legs and hydraulic accumulators. The machine is about 1m long and 0.3m wide.

Conclusions

Perhaps the most important lesson which may be learned from these projects is that legged robots will take very diverse forms in the future. We may eventually hope to see adaptive rough-ground vehicles; slow but massive underwater surveying and construction machinery; fast and agile running machines; specialized mechanisms for crawling over pressure vessels and building structures in space; and a variety of lower performance and rather fanciful forms for education and entertainment. These applications are discussed in the next chapter.

Chapter 7
Applications of walking machines

The applications of walking machines can be divided into those for which legs are an alternative to other forms of locomotion and those for which they are essential. The former category includes many tasks currently considered suitable for wheeled, tracked or ground-effect vehicles (or even aircraft) but for which legs may offer advantages. These vehicles may have a human driver or may be mobile robots. The second group involves modelling legged animals.

Legged machines have hardly begun to be applied to practical problems, so this chapter deals more with potential applications than with established ones. Some may not materialize; other, unforeseen, ones may emerge. A list of uses which have been suggested will be given shortly, but first it may be noted that a legged machine can be classified according to the kinds of function it is to provide. The main categories are:

1) transport on rough ground,
2) transport in buildings (in particular, on stairs),
3) transport in unusual environments such as pipes or orbiting structures,
4) animal modelling.

These categories are not mutually exclusive. They also cut across the areas of application: for example, rough-ground transport is needed by more than one industry; while some applications require more than one function, or operation in more than one environment. A leg prosthesis, to take an extreme case, should ideally model the human legs while providing transport both on rough ground and in buildings. As with more conventional vehicles, there is a wide spread of uses — industrial, military, public service and others:

1) military transport,
2) mining,
3) nuclear engineering and other applications of remote handling,
4) orthotic and prosthetic aids,
5) planetary exploration,
6) construction and related activities,

7) agriculture and forestry,
8) artificial intelligence research,
9) firefighting and rescue,
10) marginal applications,
11) the study of locomotion (zoology and engineering),
12) education, art and entertainment.

Military Transport

The limitations of existing vehicles when on soft ground or rough terrain have been acknowledged ever since they displaced horses and mules. The source of the most sustained interest in this application has been the US Army, which has supported, at least intermittently, research on walking machines since the 1960s. At that time they formulated a specification for a 'platoon supply system' intended to match the mobility of a foot soldier (Morrison 1968):

1) The vehicle shall climb or walk cross a 30° dry slope without overturning.
2) The vehicle shall climb a 35° slope of loose sand or gravel.
3) The vehicle footprint loading shall be less than 0.3 bar on soft surfaces.
4) The vehicle shall climb 0.25m vertical obstacles at a 45° angle of approach.
5) The vehicle shall operate in soft mud or sand and gravel that is penetrated to a depth of 0.15m.
6) The vehicle turning radius shall not exceed 2m.
7) The vehicle shall negotiate any of the following conditions at a speed up to 5km/h: rocky boulder-strewn fields including vertical steps of up to 0.3m, surfaces which are slippery due to mud, snow, ice or foliage, surfaces with low bearing strength which would permit sinkage of up to 0.25m with surface loadings of 0.3 bar, inclined surfaces of up to 45° where foot soldier traction is good and up to 30° where foot soldier traction is minimal.
8) The vehicle must operate in a self-propelled mode in water too deep to wade.
9) The vehicle must enter or leave water from a 30° slope of wet clay.
10) The vehicle must pass between two trees 1.3m apart or walk upon a trail that is 1.3m wide.

An early attempt to meet this requirement was the Iron Mule Train described in Chapter 1. In retrospect it is clear that such a fixed-cycle machine is far too inflexible to cope with this specification.

Current support for work relevant to military transport takes the form of Defense Advanced Research Projects Agency (DARPA) funding for the research of centres such as Ohio State University (OSU) and Carnegie-Mellon University. In particular, OSU's Adaptive Suspension Vehicle (see Chapter 6) is a prototype rough-country vehicle intended to be within sight of the performance required of a military vehicle. It will not be able to meet all the requirements listed above for a 'mule-performance' vehicle: indeed, such performance seems a long way off. The role of the Adaptive Suspension Vehicle is, rather, that of a walking truck (eg the 1968 General Electric quadruped). It can be seen that there are several categories of military transport for which legged machines could be considered. Research has so far concentrated on infantry support and large vehicles like trucks but with enhanced mobility. Legged vehicles might also fulfil some of the roles of helicopters.

Mining

Open-cast mining (strip mining) has used giant legged draglines and shovels for decades — some examples are described by Hutchinson (1967) — although as remarked elsewhere, their walking mechanism is specialized and of limited relevance to more flexible kinds of walking. However, it may be that in future the design of these excavators will take into account recent progress in legged locomotion research, to enable them to operate on rougher ground and steeper slopes.

Deep mining mobile robots, usually remotely controlled rather than autonomous, have been considered for many years. Since mine tunnels have a fairly smooth floor, wheels or tracks are usually suitable (Thring 1983). However, they are less adequate for steeply sloping workings, where legs may be better. In some cases the machine will have to brace itself against the floor and roof, as tunnelling machines do.

In Britain the National Coal Board has studied the idea of underground robots, but much of the coal is found in thick, nearly horizontal seams lending themselves to other automated mining methods so robots have never seemed economical. This may not continue to be so for all minerals everywhere.

Nuclear Engineering and Other Cases of Remote Handling

Fixed telechiric (remotely controlled) manipulators, or teleoperators as they are sometimes known, and industrial robots are common for handling dangerous substances where the operation takes place in a small area. Examples are the handling of munitions during their manufacture, and laboratory work on radioactive materials. Teleoperators mounted on remotely controlled vehicles are used when greater distances must be covered. One example is bomb disposal; but by far the most extensive use of such machines is likely to be in nuclear engineering. This includes nuclear weapons manufacture, but a larger industry will be increasingly concerned with the radioactivity implications of nuclear power. There are several application areas within nuclear engineering, such as fuel processing, reactor maintenance, reactor decommissioning, transport of materials, waste handling and emergency action. As waste accumulates, and as the first commercial reactors near the end of their lives, the need for effective tools grows more pressing. The subject is attracting increasing attention and a book specifically on robotics in nuclear engineering has just been published (Larcombe and Halsall 1984).

So far, mobile robots for remote handling have all used wheels or tracks. There is also a race of hybrid machines with combinations of wheels and legs (see Chapter 1). Proper walking machines have not yet found much favour, probably because the geometrical shapes of surfaces within power stations encourage the use of simpler alternatives. An exception is the magnetically clinging inspection robot mentioned in Chapter 6. However, this may well change: in the long run the versatility of a legged robot will give it advantages in some situations.

Prosthetics and Orthotics; Walking Chairs

One of the first uses to be recognized for walking machines was as a prosthesis for patients with both legs amputated, or as an orthotic aid for paraplegics. Much of the research on bipedal locomotion originates in this application. A successful device would have to be able to negotiate stairs as well as level and sloping ground.

Several research projects on prosthetic and orthotic aids are described in Chapter 1, although a proper treatment of this subject is outside the scope of the book. The current research on biped locomotion (see Chapter 6) may also lead to aids of this kind.

An alternative to a bipedal prosthesis is a legged equivalent of a wheelchair, using four or more legs. This is in principle easier as there is less of a problem with stability. Attempts were made to design walking chairs in the era of spoked wheels and purely mechanical linkages (Vukobratovic 1973; Thring 1983) but were not very successful. The greatest problem is turning a corner on a staircase. A secondary problem is simply making the whole machine compact enough to manoeuvre in the confined spaces of a house. This problem is particularly acute if it is to climb stairs, keeping its occupant more or less level and the centre of gravity of the whole within the base of support. Nevertheless, it appears soluble with modern methods. Probably the easiest approach is that of Hitachi's wheel-leg hybrid described in Chapter 1, although this is not a pure walking machine.

Planetary Exploration

Space travel and robots were twin pillars of science fiction long before there was any realistic prospect of either. There is, of course, a sound reason for using robots for space and planetary exploration, namely the desirability (sometimes) of dispensing with human passengers. Perhaps the most interesting aspect of space robotics is that the problems of communication confer an unusually high value on autonomous intelligence, so that a Mars rover, for example, is a more valid object of artificial intelligence research than are many Earth-bound robots.

Turning to legged machines specifically, as on Earth, legged locomotion must be considered as an alternative to other forms; also, it may be applied to manned or unmanned vehicles. Legged machines have been proposed (Seifert 1967) but never built; after all, only two land vehicles of any kind have ever been operated off the Earth (both wheeled designs) so it is too early to expect rapid progress. Legged vehicles would probably offer better mobility on the loose, steep and rocky surfaces common on the Moon and Mars, but at present their complexity and poor ratio of payload to structural weight must count against them.

Construction and Related Activities

Construction work on land, under water and in space can be included here. Of these, only in underwater operations has there so far been a demonstration of the practical capabilities of walking machines. This is the use of the Komatsu octopod robot ReCUS for sea-bed civil engineering work. The machine is described briefly in Chapter 6 (see also Ishino *et al.* 1983). So far it has been used for surveying the sea-bed in preparation for building the piers of the Honshu-Shikoku bridge. The survey is both photographic, using television and still cameras looking downwards through a clear viewing device (a cone containing clear water), and ultrasonic to give a topographic profile.

Legs were preferred to tracks because of the need to provide a stable, level platform on surfaces with a roughness of up to 2m and in tidal currents of up to 6 knots. A tracked vehicle would have had to be enormous, and carry a platform levelling mechanism.

ReCUS is being considered for other work. A good candidate is levelling the rubble used in breakwater foundations. A blade or rake and a roller would slide along the main frame and could be kept accurately horizontal, unlike that of a tracked bulldozer. Other proposed applications are: an underwater excavator with rippers or rotary cutters, an underwater drilling machine, a pipeline trench digger, and a handling robot with manipulators for work with underwater structures.

Turning to construction on land, walking machines have not yet been used. For trucks and excavators to be used on building sites a potential advantage of legs is the ability to step over obstacles such as trenches and delicate structures such as partly built walls and kerbs. Also easily damaged by conventional vehicles are soil and vegetation. Much damage is often done by cutting access roads, which unless carefully surfaced can degenerate into deep mud. It might even be possible to open up some sites which are inaccessible by ordinary vehicles.

Agriculture and Forestry

Robotics is entering agriculture more through attempts to automate the picking and handling of crops than in the form of developments in locomotion. Attempts have been made to automate ploughing and similar operations using wheeled tractors. For example, in Britain the National Institute of Agricultural Engineering has done some work on guiding ploughing tractors using optical sensing of reflective posts on the edge of the field (Harries and Ambler 1981). As far as legged locomotion is concerned no serious work has been done. Thring (1983) has done some preliminary work on designing legged tractors using all-mechanical methods of driving the legs although, as explained earlier, a purely mechanical system is unlikely to be adequate.

There are two obvious advantages of a legged agricultural machine over a wheeled one. First, it should produce less damage to the ground and to crops; and second, if the stride and ground clearance are large enough, it might be possible to step over fences and other obstacles. It is also possible that legged machines would be better in deep snow and mud.

A possible use in forestry is timber extraction. Machines already exist which use wheels to carry most of the weight but legs for additional support and propulsion. They can operate on steep slopes but their mobility is still limited. A machine relying entirely on legs should be better at negotiating obstacles such as fallen logs. A second use would be to patrol forests to detect and report fires, although the need for legs is less clear.

Mobile Robots in Artificial Intelligence

Since the early days of artificial intelligence in the 1960s, robots have been seen as a tool for testing theories and programs against the real world. One of the difficulties with artificial intelligence is that theories which look plausible, and work when tested on simple problems, often break down when confronted with a real-world task. Robots are useful for testing theories because they force the designer to deal with the complexity and untidiness of the real world. Specifically, a robot forces the software designer to consider:

1) the problems of perception with noisy signals, defective sensors and extremely complex information sources such as vision,
2) how to represent the world and update the representation as the robot moves and its environment changes,
3) how to plan actions to achieve goals,
4) the execution of plans and the monitoring and scheduling of behaviour,
5) the problems of using actuators which do not produce exactly the intended motion.

A well-known example was Stanford Research Institute's Shakey (Raphael 1976); another research robot was Berkeley's JASON (Coles *et al.* 1975), and a series of them have been built at Queen Mary College, London. The legged WABOT, described in Chapter 1, also perhaps falls into this category.

These seeing, planning, do-everything robots have gone out of fashion slightly at present, as the individual components listed above have split off into sub-fields of artificial intelligence; they have all turned out to be larger and more difficult subjects than was thought in the early days. However, it is only a matter of time before a new generation of much better, intelligent robots emerges, and when it does some of them will probably have legs.

Fire-fighting and Rescue Robots

Mobile robots have been proposed for rescuing people in fires (Kobayashi and Nakamura 1983) and for fire-fighting (Thring 1983). Thring built a wheeled robot to detect and attack a fire, using heat and light sensors. Fire or security patrol robots have been suggested from time to time (for example, Russell 1983) but so far little work has been done.

For a fire-fighting robot to be of much use in buildings it would need to climb stairs, so legs are desirable. The problem is perhaps easier than some stair-climbing applications because in an emergency a lurching motion and damage to furnishings would be more acceptable. Legged vehicles might also be desirable for mountain rescue, as they offer the possibility of relatively smooth movement on rough paths unusable by normal vehicles.

Applications Unlikely to be Suitable for Walking Machines

There are some applications of mobile robots where it is hard to see any advantages in legs. However, conditions may change, so it is worth mentioning them. Perhaps the most common is the automatically guided vehicle.

This term is usually reserved for wheeled vehicles which use a buried wire or other guidance method to follow a fixed path. Their most common use is in factories and warehouses where they can replace manned fork-lift trucks or conveyors. A well-known example is their use in carrying partially built car bodies from one welding station to another. In this application, the fleet of robotic transporters replaces the normal straight track of a production line.

Current research on automatically guided vehicles aims to endow them with more intelligent behaviour on meeting obstacles. A second aim is to allow them to navigate without buried wires or other inconvenient installations. These aims imply the need to sense the environment by vision, sonar, touch and other means (one centre of such research is Warwick University; see Larcombe 1979).

It is unlikely that walking vehicles will replace wheeled ones in most automatically guided vehicle applications, as the factories where they are used are designed for wheels. However, some of the control and navigation issues are relevant to less constrained robots.

The Study of Locomotion (Zoology and Engineering)

Research on animal locomotion aims to discover such things as why animals use particular gaits, relationships between body size, speed, oxygen consumption, efficiency and so on. Another aim is to reconstruct the behaviour and physiology of extinct animals from fossil remains.

Many of these aims are concerned with features specific to biological systems, for example muscle, and so must use actual animals. However, work with animals has many practical problems, and also often involves cruel and macabre experiments. An example is J.T. Manter's research (Manter 1938) on the dynamics of quadrupedal walking. This involved training a cat to walk over a series of force platforms. The cat was then killed, frozen and dismembered. The moment of inertia of a component such as a paw was determined by sticking a wire through it, swinging it as a pendulum and timing the oscillation.

It seems possible that legged robots will eventually be sufficiently advanced to allow some characteristics of locomotion to be studied using machines instead of animals. For example, it is easy to scale a machine up, or change the mass of certain parts, and it can be made to repeat the same behaviour (eg a particular gait) reliably. Thus, it might be possible to study behaviours which animals are capable of but unwilling to use. There are precedents for this approach, notably in the study of flight. Replicas of pterosaurs have been built to investigate the flight characteristics of different reconstructions. So far these replicas have been passive gliders or wind-tunnel models, but it has recently been proposed that a full-scale powered pterosaur be built (Padian 1984), which would fly under radio control.

The study of legged locomotion in its own right need not be confined to reproducing that of animals. One of the aims of research such as that on one-legged hopping machines is to explore the possibilities of locomotion with legs, unconstrained by what biological evolution has selected.

Education, Art and Entertainment

Research on the reconstruction of extinct animals is done largely by museums (the Smithsonian is involved in the pterosaur study, discussed earlier), and is associated with their role of public exhibition. This suggests that walking models of animals such as dinosaurs could be built for exhibition; they would fit in with the tradition of working models demonstrating biological structure and function, although calling for some innovations in display methods. To take this approach further, it is possible to envisage a zoo or safari park setting for free-ranging animal models, especially the larger ones. There seems to be no technical reason why full-scale walking models of some of the sauropods, for example, could not be built. It is probably unreasonable to expect verisimilitude of both appearance and movement, at least in early attempts, and the models may best be left frankly mechanical where the demonstration of movement is important, rather than being forced into an unconvincing rubbery habit.

Exhibits of this kind play a multiple role: they can be both educational and entertaining. They could also be built commercially purely as entertainments for amusement parks. For this function there is no need to adhere to zoologically plausible forms: fantastic walking machines like those seen in science-fiction films could be built. This is probably easier than trying to make an accurate animal model.

Finally, it is worth mentioning that there is clearly a large sculptural element in a machine designed to imitate an animal or a science-fiction robot. Some legged robots may even come to be designed purely to express aesthetic qualities, of form and movement. Thus, the subject of walking machines may become a branch both of engineering and of art.

References and bibliography

Notes and Abbreviations

A second year in square brackets is the year of the conference if this was earlier than the year of publication. Abbreviations of conference series and journal titles:

CISM: Centre International des Sciences Mécaniques
ICAR: International Conference on Advanced Robotics
IFAC: International Federation of Automatic Control
IFToMM: International Federation for the Theory of Machines and Mechanisms
IJCAI: International Joint Conference on Artificial Intelligence
IJRR: International Journal of Robotics Research
ISIR: International Symposium on Industrial Robots
ISTVS: International Society for Terrain Vehicle Systems
PWN: Polish Scientific Publishers, Warsaw

References

Alexander, R.McN. Optimum walking techniques for quadrupeds and bipeds. *J. Zool. (Lond)* 1980, **192**, 97-117.

Alexander, R.McN. *Locomotion of Animals* Blackie, Glasgow, 1982.

Alexander, R.McN. The gaits of bipedal and quadrupedal animals. *IJRR* 1984, **3** (2 – Summer), 49-59.

Bardelli, R.; Dario, P.; DeRossi, D.; Pinotti, P.C. Piezo- and pyroelectric polymers: skin-like tactile sensors for robots and prostheses. In *13th ISIR* April 1983.

Bavarian, B.; Wyman, B.F.; Hemami, H. Control of the constrained planar simple inverted pendulum. *Int. J. Control* 1983, **37** (4), 741-753.

Bekker, M.G. *Theory of Land Locomotion* University of Michigan Press, Ann Arbor, 1956.

Bekker, M.G. *Introduction to Terrain-Vehicle Systems* University of Michigan Press, Ann Arbor, 1969.

Beletskii, V.V. Motion control of biped walking robots. *8th IFAC Symp. on Automatic Control in Space* 1979, pp. 317-322.

Benjamin, H.L. The development of a production robot tactile position sensor. In *13th ISIR* April 1983.

Bessonov, A.P.; Umnov, N.V. The analysis of gaits in six-legged vehicles according to their static stability. In *1st CISM-IFToMM Symp. on Theory & Practice of Robots & Manipulators* 1973.

Bessonov, A.P.; Umnov, N.V. Choice of geometric parameters of walking machines. In *2nd CISM-IFToMM Symp. on Theory & Practice of Robots & Manipulators* Elsevier, 1977 [1976].

Bessonov, A.P.; Umnov, N.V. Features of kinematics of turn of walking vehicles. *3rd CISM-IFToMM Symp. on Theory & Practice of Robots & Manipulators* Elsevier, 1980, pp. 87-97.

Bessonov, A.P.; Umnov, N.V. The stabilization of the position of the body of walking machines. *Mechanism & Machine Theory* 1983, **18** (4), 261-265.

Coles, L.S.; Robb, A.M.; Sinclair, P.L.; Smith, M.H.; Sobek, R.R. Decision analysis for an experimental robot with unreliable sensors. In *IJCAI-75* 1975.

Dario, P.; Domenici, C.; Bardelli, R.; DeRossi, D.; Pinotti, P.C. Piezoelectric polymers: new sensor materials for robotic applications. In *13th ISIR* April 1983.

Devjanin, E.A.; Gurfinkel, V.S.; Gurfinkel, E.V.; Kartashev, V.A.; Lensky, A.V.; Shneider, A.Yu.; Shtilman, L.G. A six-legged walking robot capable of terrain adaptation. *Mechanism & Machine Theory* 1983, **18** (4), 257-260.

Donner, M.D. The design of OWL: a language for walking. *ACM SIGPLAN Notices* 1983, **18** (6), 158-165.

Gabrielli, G.; von Karmen, T.H. What price speed? *Mech. Eng.* 1950, **72** (10), 775-781.

Gaillet, A.; Reboulet, C. An isostatic six component force and torque sensor. In *13th ISIR* April 1983.

Gardner, J.F. *et al.* Design and testing of a digitally controlled hydraulic actuation system for a walking vehicle leg mechanism. *Proc. 8th Appl. Mechanisms Conf.* Oklahoma State University, St Louis, Missouri, 1983, pp. 2-1 to 2-7.

Gordon, J.E. *Structures* Penguin Books Ltd, 1978.

Grundman, J.; Seireg, A. Computer control of a multi-task exoskeleton for paraplegics. *2nd CISM-IFToMM Symp. on Theory & Practice of Robots & Manipulators* Elsevier, 1977 [1976], pp. 233-240.

Gubina, F.; Hemami, H,; McGhee, R.B. On the stability of biped locomotion. *IEEE Trans. Biomed. Eng.* 2 March 1974, **BME-21**, 102-108.

Hall, J.I.; Witt, D.C. The development of an automatically stabilised powered walking device. In *I. Mech. E. Conf. on Human Locomotor Engineering* 1971.

Harries, G.O.; Ambler, B. Automatic ploughing: a tractor guidance system using opto-electronic sensing and a microprocessor based controller. *J. Agric. Engineering Res.* 1981, **26**, 33-53.

Hemami, H. Reduced order models for biped locomotion. *Proc. 7th Pittsburgh Conf. on Modeling & Simulation* 1976, pp. 270-276.

Hemami, H.; Chen, B. Stability analysis and input design of a two-link planar biped. *IJRR* 1984, **3** (2), 93-100.

Hemami, H.; Jeswa, V.C.; McGhee, R.B. Some alternative formulations of manipulator dynamics for computer simulation studies. In *Proc. 13th Allerton Conf. on Circuit & System Theory* 1975.

Hemami, H.; Katbab, A. Constrained inverted pendulum model for evaluating upright postural stability. *J. Dyn. Syst. Measurement Contr.* 1982, **104** (4), 343-349

Hemami, H.; Tomovic, R.; Ceranowicz, A.Z. Finite state control of planar bipeds with application to walking and sitting. *J. Bioengineering* 1979, **2**, 477-494.

Herreid, C.F.; Fourtner, C.R. *Locomotion and Energetics in Arthropods* Plenum Press.

Higdon, D.I.; Canon, R.H. On the control of unstable multiple-output mechanical systems. *Proc. Winter Annual Meeting* American Society of Mechanical Engineers, New York, 1963.

Hildebrand, M. *Analysis of Vertebrate Structure* John Wiley & Sons, 1974.

Hildebrand, M. Analysis of tetrapod gaits. *Neural Control of Locomotion* eds R.M. Herman *et al.*, Plenum Press, 1976, pp. 203-236.

Hirose, S. A study of the design and control of a quaduped walking vehicle. *IJRR* 1984, 3 (2 — Summer) 113, 133.

Hirose, S.; Masui, T.; Kikuchi, H.; Fukuda, Y.; Umetani, Y. Titan III: a quadruped walking vehicle. *2nd Int. Symp. of Robotics Research* August 1984, Kyoto, Japan, pp. 247-253.

Hirose, S.; Umetani, Y. Kinematic control of an active cord mechanism with tactile sensors. *2nd CISM-IFToMM Symp. on Theory & Practice of Robots & Manipulators* Elsevier, 1977 [1976], pp. 241-252.

Hirose, S.; Umetani, Y. Some considerations on a feasible walking mechanism as a terrain vehicle. *3rd CISM-IFToMM Symp. on Theory & Practice of Robots & Manipulators* Udine, Italy, 1978 (published by PWN, Warsaw, 1980).

Hirose, S.; Umetani, Y.; Oda, S. An active cord mechanism with oblique swivel joints, and its control. *4th CISM-IFToMM Symp. on Theory & Practice of Robots Manipulators* PWN, Warsaw, 1983 [1981], pp. 327-340.

Hutchinson, A.C. Machines can walk. *The Chartered Mechanical Engineer* November 1967, pp.480-484.

Ichikawa, Y.; Ozaki, N.; Sadakane, K. Five-legged vehicle for remote systems in nuclear power plants. In *Proc. 30th Conf. Remote Syst. Technol.* 1982.

Ichikawa, Y.; Ozaki, N.; Sadakane, K. A hybrid locomotion vehicle for nuclear power plants. In *IEEE Trans. Systems, Man & Cybernetics* December 1983, **SMC-13** (6).

Ishino, Y.; Naruse, T.; Sawano, T.; Honma, N. Walking robot for underwater construction. *ICAR* 1983, pp. 107-114.

Iwamoto, T.; Yamamoto, H.; Honma, N. Transformable crawler mechanism with adaptability to terrain variations. *ICAR* 1983, pp. 285-291.

Kato, T. *et al.* Information-power machine with senses and limbs. In *1st CISM-IFToMM Symp. on Theory & Practice of Robots & Manipulators* Springer-Verlag, 1974 [1973].

Kato, T.; Takanishi, A.; Ishikawa, H.; Kato, I. The realization of quasidynamic walking by a biped walking machine. *4th CISM-IFToMM Symp. on Theory & Practice of Robots & Manipulators* PWN, Warsaw, 1983 [1981], pp. 341-351.

Kessis, J.J.; Rambaut, J.P.; Penne, J. Walking robot multi-level architecture and implementation. *4th CISM-IFToMM Symp. on Theory & Practice of Robots & Manipulators* PWN, Warsaw, 1983 [1981], pp. 297-304.

Klein, C.A.; Olson, K.W.; Pugh, D.R. Use of force and attitude sensors for locomotion of a legged vehicle over irregular terrain. *IJRR* 1983, 2 (2 — Summer), 3-17.

Klein, C.A.; Wahawisan, W. Use of a multiprocessor for control of a robotic system. *IJRR* 1982, 1 (2 — Summer), 45-59.

Kemmochi, S.; Kazuoka, S. Mechanized devices for the inservice inspection of nuclear power plants. *ICAR* 1983, pp. 83-90.

Kobayashi, A.; Nakamura, K. Rescue robot for fire hazards. *ICAR* 1983, pp. 91-106.

Kohler, G.; Selig, M.; Salaske, M. In *Proc. 24th Conf. on Remote Systems Technology* American Nuclear Society, 1976.

Kugushev, E.I.; Jaroshevskij, V.S. Problems of selecting a gait for an integrated locomotion robot. In *IJCAI-75* 1975.

Larcombe, M.H.E. Mobile robots for industrial use. In *Robots '79* conference of the British Robot Association, 1979.

Larcombe, M.H.E.; Halsall, J.R. *Robotics in Nuclear Engineering* Graham & Trotman, 1984.

Liston, R.A. Walking machines. *Terramechanics Journal* 1964, 1 (3), 18-31.

Liston, R.A. *Walking Machine Studies* US Army Tank Automotive Center, Warren, Michigan, 1966.

Lozano-Perez, T. *Robot Programming* Massachusetts Institute of Technology AI Memo number 698, December 1982.

Manter, J.T. The dynamics of quadrupedal walking. *J. Expl. Biol.* 1938, 15, 522-540.

Matsuoka, K. A mechanical model of repetitive hopping movements. *Biomechan.* 1980, 5, 251-258 (in Japanese).

Mavaddat, F. WATSON/1: WATerloo's SONically guided robot. *J. Microcomputer Applications* 1983, 6, 37-45.

McCloy, D.; Martin, H.R. *Control of Fluid Power* Ellis Horwood, 1980.

McGhee, R.B. Some finite state aspects of legged locomotion. *Math. Biosci.* 1968, 2 (1), 67-84.

McGhee, R.B. Robot locomotion. In *Neural Control of Locomotion* eds Herman *et al.*, Plenum, 1976, pp. 237-264.

McGhee, R.B. Control of legged locomotion systems. *Proc. 18th Automatic Control Conf.* San Francisco, June 1977, pp. 205-215.

McGhee, R.B.; Iswandhi, G.I. Adaptive locomotion of a multilegged robot over rough terrain. *IEEE Trans. Systems: Man & Cybernetics* 1979 [1978], **SMC9** (4), 176-182.

McMahon, T.A. Mechanics of locomotion. *IJRR* 1984, 3 (2 − Summer), 4-28.

Miura, H.; Shimoyana, I. Dynamic walk of a biped. *IJRR* 1984, 3 (2 − Summer), 60-74.

Miyazaki, F.; Arimoto, S. A design method of control for a biped walking machine. *4th CISM-IFToMM Symp. on Robots & Manipulators* PWN, Warsaw, 1983a [1981], pp. 317-326.

Miyazaki, F.; Arimoto, S. A hierarchical control for biped robots. *ICAR* 1983b pp. 299-306.

Mizen, N.J. Machines with strength. *Science Journal* 10 October 1968, 4, 50-55.

Morrison, R.A. Iron mule train. In *Cornell Aeronautical Lab/ISTVS Off-road Mobility Research Symposium* Washington DC, June 1968.

Mosher, R.S. Test and evaluation of a walking truck. In *Cornell Aeronautical Lab/ ISTVS Off-road Mobility Research Symposium* Washington DC, June 1968.

Nichols, G.K.; Witt, D.C. An experimental unpowered walking aid. *Engineering in Medicine* 1 October 1971, pp. 7-11.

Ogo, K.; Ganse, A.; Kato, I. Quasidynamic walking of biped walking machine aiming at completion of steady walking. In *3rd CISM-IFToMM Symp. on Robots & Manipulators* 1980 [September 1978].

Okhotsimski, D.E. Motion control system for a mobile robot. *8th IFAC Symp. on Automatic Control in Space* 1979, pp. 251-256.

Orin, D.E. Supervisory control of a multilegged robot. *IJRR* 1982, 1 (1 − Spring), 79-91.

Orin, D.E.; McGhee, R.B. Dynamic computer simulation of robotic mechanisms. *4th CISM-IFToMM Symp. on Robots & Manipulators* 1983 [September 1981], pp. 286-296.

Orin, D.E.; McGhee, R.B.; Vukobratovic, M.; Hartoch, G. Kinematic and kinetic analysis of open-chain linkages utilizing Newton-Euler methods. *Math. Biosciences* 1979, 43, 107-130.

Orin, D.E.; Oh, Y. A mathematical approach to the problem of force distribution in locomotion and manipulation systems containing closed kinematic chains. In *3rd CISM-IFToMM Symp. on Robots & Manipulators* 1980 [September 1978].

Ozaki, N.; Suzuki, M.; Ichikawa, Y. Tele-operated mobile robot for remote maintenance in nuclear facilities. *ICAR* 1983, pp. 67-74.

Ozguner, F.; Tsai, S.J.; McGhee, R.B. An approach to the use of terrain-preview information in rough-terrain locomotion by a hexapod walking machine. *IJRR* 1984, 3 (2 — Summer), 134-146.

Padian, K. Flight of fancy planned for the largest pterosaur. *Nature (London)* 11 October 1984, **311**, 511.

Park, W.T.; Fegley, K.A. In *5th IFAC Symp. on Automatic Control in Space* Genoa, 1973.

Pearson, K.G.; Franklin, R. Characteristics of leg movements and patterns of coordination in locusts walking on rough terrain. *IJRR* 1984, 3 (2 — Summer), 101-112.

Petternella, M.A.D.; Salinari, S.A.M. Simulation by digital computer of a walking machine control system. In *5th IFAC Symp. on Automatic Control in Space* 1973.

Platonov, A.K. Problems of motion control for mobile robots. *8th IFAC Symp. on Automatic Control in Space* 1979, pp. 303-310.

Pollock, N. Electronic compass using a fluxgate sensor. *Wireless World* October 1982, pp. 49-54.

Pollock, N. Low-cost servo accelerometer. *Wireless World* May 1983, pp. 66-70.

Raibert, M.H.; Brown, H.B.; Chepponis, M.; Hastings, E.; Murthy, S.S.; Wimberley, F.C. *Dynamically Stable Legged Locomotion: 2nd Report to DARPA* October 1981 - December 1982, The Robotics Institute, Carnegie-Mellon University, Pittsburgh, 27 January 1983 (a).

Raibert, M.H.; Brown, H.B.; Chepponis, M.; Hastings, E.; Koechling, J.; Murthy, K.N.; Murthy, S.S.; Stentz, A.J. *Dynamically Stable Legged Locomotion: Progress Report* October 1982 - October 1983 (CMU-RI-TR-83-20) The Robotics Institute, Carnegie-Mellon University, Pittsburgh, 13 December 1983(b).

Raibert, M.H.; Brown, H.B.; Chepponis, M. Experiments in balance with a 3D one-legged hopping machine. *IJRR* 1984, 3 (2 — Summer), 75-92.

Raibert, M.H.; Sutherland, I.E. Machines that walk. *Scientific American* 1 January 1983, **248**, pp. 32-41.

Raphael, B. *The Thinking Computer* Freeman, 1976.

Reichardt, J. *Robots: Fact, Fiction & Prediction* Thames and Hudson, 1978.

Russell, M. *ODEX I: the first functionoid. Robotics Age* September/October 1983, 5 (5), pp. 12-18.

Seifert, H.S. The lunar pogo stick. *J. Spacecraft & Rockets* 1967, 4 (7), 941-943.

Shen, C.N.; Kim, C.S. A laser rangefinder path selection system for martian rover using logarithmic scanning scheme *Proc. 8th IFAC Symp. on Automatic Control in Space* Pergamon, Oxford, 1980, pp. 323-330.

Shigley, J.E. The mechanics of walking vehicles. In *Proc. 1st Int. Conf. on Mechanics of Soil-Vehicle Systems* Turin, 1961.

Stepanenko, Yu.; Vukobratovic, M. Dynamics of articulated open-chain active mechanisms. *Math. Biosciences* 1976, **28** (1/2), 137-170.

Sutherland, I.E.; Ullner, M.K. Footprints in the asphalt. *IJRR* 1984, 3 (2 — Summer), 29-36.

Taguchi, K.; Ikeda, K.; Matsumoto, S. Four-legged walking machine *2nd CISM-IFToMM Symp. on Theory & Practice of Robots & Manipulators* PWN, Warsaw, 1977 [1976], pp. 163-171.

Thring, M.W. *Robots and Telechirs* Ellis Horwood, 1983.

Todd, D.J. An experimental study of pneumatic walking robots. In *Digital Systems for Industrial Automation* 1984, **2** (4).

Tomovic, R. A general theory of creeping displacement. *Cybernetics* 1961, **4** (2), 98-107.

Tomovic, R.; McGhee, R.B. A finite state approach to the synthesis of bioengineering control systems. *IEEE Trans. on Human Factors in Electronics* June 1966, **HFE-7**, 65-69.

Trautwein, W. Control strategies for planetary rover motion and manipulator control. In *5th IFAC Symp. on Automatic Control in Space* 1973.

Vertut, J.; Marchal, P.; Corfa, Y.; Francois, D. Vehicles with wheels and legs: the in pipe remote inspection vehicle and his family. In *3rd CISM-IFToMM Symp. on Theory & Practice of Robots & Manipulators* September 1978, Elsevier, Oxford and PWN, Warsaw, 1980 [September 1978].

Vohnout, V.J.; Alexander, K.S.; Kinzel, G.L. The structural design of the legs for a walking vehicle. *8th Applied Mechanisms Conf.* St Louis, Missouri, 1983, pp. 50-1 to 50-8.

Vukobratovic, M. Legged locomotion robots: mathematical models, control algorithms, and realizations. In *5th IFAC Symp. on Automatic Control in Space* Genoa, 1973.

Vukobratovic, M.; Stepanenko, J. On the stability of anthropomorphic systems. *Math. Biosci.* 1972, **15**, 1-37.

Vukobratovic, M.; Hristic, D.; Stojiljkovic, Z. Development of active anthropomorphic exoskeletons. *Medical & Biological Eng.* January 1974, pp. 66-80.

Waldron, K.J.; Vohnout, V.J.; Pery, A.; McGhee, R.B. Configuration design of the adaptive suspension vehicle *IJRR* 1984, **3** (2 – Summer), 37-48.

Waldron, K.J.; Kinzel, G.L. The relationship between actuator geometry and mechanical efficiency in robots. *4th CISM-IFToMM Symp. on Theory & Practice of Robots & Manipulators* PWN, Warsaw, 1983 [September 1981], pp. 305-316.

Walker, M.W.; Orin, D.E. Efficient dynamic computer simulation of robotic mechanisms. *J. Dynamic Systems, Measurement & Control* September 1982, **104**, 205-211.

Wells, M. *Lower Animals* Weidenfeld & Nicholson, 1968.

Wilson, D.M. Insect walking. *Ann Rev. Entomol.* 1966, **11**, 103-122.

Witkowski, C.M.; Bond, A.H.; Burton, M. The design of sensors for a mobile teleoperator robot. *Digital Systems for Industrial Automation* 1983, **2** (1), 85-111.

Young, S.J. *Real Time Languages* Ellis Horwood, 1982.

Zuk, D. *3D Optical Terrain Sensor for the 1984 ASV* Workshop Summary, AF/DARPA Robotics Workshop, 1983, pp. 11-11 to 11-27.

Bibliographical Notes

TOYS AND AUTOMATA

An eclectic if uncritical compendium of robots in art, fiction, entertainment and industry is Reichart (1978). This gives several references to the history of toy robots and automata, such as Chapuis and Droz (1958) and Hillier (1976).

ROBOT DYNAMICS AND CONTROL

Several publishers are producing series of titles in robotics. They deal primarily with manipulators, but some aspects are relevant to legged robots. One such series is that by Springer-Verlag on *Scientific Fundamentals of Robotics*, of which the first two members are Vukobratovic and Potkonjak (1982) and Vukobratovic and Stokic (1982). A second series, covering engineering as well as mathematical aspects of robotics, is Kogan Page's *Robot Technology Series*. Some books in this series which are relevant to legged robots are Coiffet (1983a and b), Lhote *et al.* (1984), Parent and Laurgeau (1984) and Vertut and Coiffet (1984). Two books on the theory of robot dynamics and control by MIT Press are Brady *et al.* (1983) and Paul (1981).

Brady, M.; Hollerbach, J.M.; Johnson, T.L.; Lozano-Perez, T.; Mason, M.T. *Robot Motion: Planning and Control* MIT Press, Massachusetts, 1983.

Chapuis, A.; Droz, E. *Automata — a Historical and Technological Study* Neuchatel, 1958.

Coiffet, P. *Modelling and Control* Kogan Page, London, 1983.

Coiffet, P. *Interaction with the Environment: Robot Sensors and Sensing* Kogan Page, London, 1983.

Hillier, M. *Automata and Mechanical Toys* London, 1976.

Lhote, F.; Kauffmann, J-M.; André, P.; Taillard, J-P. *Robot Components and Systems* Kogan Page, London, 1984.

Parent, M.; Laurgeau, C. *Logic and Programming* Kogan Page, London, 1984.

Paul, R.P. *Robot Manipulators: Mathematics, Programming and Control* MIT Press, Massachusetts, 1981.

Vertut, J.; Coiffet, P. *Teleoperations and Robotics: Evolution and Development* Kogan Page, London, 1985.

Vertut, J.; Coiffet, P. *Teleoperations and Robotics: Applications and Technology* Kogan Page, London, 1985.

Vukobratovic, M.; Potkonjak, V. *Dynamics of Manipulation Robots* Springer-Verlag, Berlin, 1982.

Vukobratovic, M.; Stokic, D. *Control of Manipulation Robots* Springer-Verlag, Berlin, 1982.

Index